Fair Trade and Organic Initiatives in Asian Agriculture

In addition to constituting an evolving area of inquiry within the social sciences, agricultural certification, and particularly its Fair Trade and organic components, has emerged as a significant tool for promoting rural development in the global South. This book is unique for two reasons. First, in contrast to existing studies that have tended to examine Fair Trade and organic certification as independent systems, the studies presented in this book reveal their joint application within actual production settings, demonstrating the greater complexity entailed in these double certification systems through the generation of contradictions and tensions compared with single certification systems. Second, the authors, who are both Asian, reveal the realities of applying Fair Trade and organic certification systems within Asian agriculture. In doing so, they challenge the fact that most Fair Trade studies have been undertaken by Western scholars who have tended to focus on Latin American and African producers. Drawing on a wealth of grounded case studies conducted in India, Thailand, and the Philippines, this pioneering study on double certification makes a significant contribution to studies on Fair Trade and organic agriculture beyond Asia.

Rie Makita is a Professor in the Faculty of International Social Sciences at Gakushuin University, Japan. Her publications include refereed articles on fair trade and organic agriculture in *Development in Practice*, *Development Policy Review*, *Geoforum*, and *Asia Pacific Viewpoint*.

Tadasu Tsuruta is a Professor in the Faculty of Agriculture at Kindai University, Japan. His publications include "Socioeconomic Impact of Alternative Trade on Southern Thai Villages: A Case Study on the Production of Herbicide-Free Bananas for Japanese Consumers" for *Perspective of Alternative Commodities Chain*, 2008.

Routledge Studies in Development Economics

For a complete list of titles in this series, please visit www.routledge.com/series/SE0266

Fair Trade and Organic Initiatives in Asian Agriculture

The Hidden Realities

Rie Makita and Tadasu Tsuruta

Routledge
Taylor & Francis Group

LONDON AND NEW YORK

First published 2017 by Routledge

2 Park Square, Milton Park, Abingdon, Oxfordshire OX14 4RN
52 Vanderbilt Avenue New York NY 10017

Routledge is an imprint of the Taylor & Francis Group, an informa business

First issued in paperback 2019

British Library Cataloguing-in-Publication Data
A catalogue record for this book is available from the British Library

Library of Congress Cataloging-in-Publication Data
A catalog record for this book has been requested

ISBN: 978-1-138-65314-6 (hbk)
ISBN: 978-0-367-35048-2 (pbk)

Typeset in Galliard
by Apex CoVantage, LLC

Dedicated to the memory of Mr. Edappallil M. Koshy

Contents

Figures

Tables

Acknowledgments

We are indebted to a range of institutions, as well as individuals, who have made the writing of this book possible. To begin with institutions, our research was financially supported by KAKENHI, the Japan Society for the Promotion of Science (Project Nos. 22830092, 24520900, 3402028, and 26301020) and the Integrated Research System for Sustainability Science at the University of Tokyo. Earlier versions of four chapters in this book were previously published as follows: "A confluence of Fair Trade and organic agriculture in southern India" (Chapter 2), published in 2011 in *Development in Practice*, 21(2), 205–217; "Fair Trade and organic initiatives confronted with Bt cotton in Andhra Pradesh, India: A paradox" (Chapter 3), published in 2012 in *Geoforum*, 43(6), 1232–1241; "Fair Trade certification: The case of tea plantation workers in India" (Chapter 4), published in 2012 in *Development Policy Review*, 30(1), 87–107; and "Livelihood diversification with certification-supported farming: The case of land reform beneficiaries in the Philippines" (Chapter 6), published in 2016 in *Asia Pacific Viewpoint*, 57(1), 44–59. We thank the respective journal editors for granting us permission to reproduce the above-mentioned material.

A number of individuals have aided the fieldwork conducted for the case studies reported in this book. We particularly thank George Andu K., Jiju Robinson, Stan Bonagri, Govindula Venkat Raman, Mathew Mathai, and Ramakrishna Rao in India; Suriya Chanachai, Porphant Ouyyanont, Apinya Wanaset, and Vitoon Panyakul in Thailand; and Masahiko Hotta, Edwin Lopez, Thelma Guanzon, Caren Pineda, and Makoto Ueda in the Philippines. We are, of course, deeply grateful to the distributors of Fair Trade and organic products, farmers in the study villages, and managers and workers in the study plantations, who took the time to meet with us and who patiently answered our questions. To protect their anonymity, we have not revealed their names.

Several colleagues kindly read and commented upon selected draft chapters. These individuals include Lesley Potter, Katherine Gibson, Alastair Smith, Padraig Carmody, and Japanese scholars within the Alter Trade study team led by Koichi Ikegami. We are grateful to all of these individuals for their helpful suggestions and input. Finally, we extend our appreciation to Yong Ling Lam and Samantha Phua, who facilitated the publication process.

Abbreviations

AAN   Alternative Agriculture Network
ACT   Organic Agriculture Certification Thailand
ADB   Asian Development Bank
AOFG   Agriculture and Organic Farming Group India
ATC   Alter Trade Corporation
ATFI   Alter Trade Foundation Incorporation
Bt   *Bacillus thuringiesnsis*
CAEF   Community of Agro-Ecology Foundation
CARP   Comprehensive Agrarian Reform Program
CHAI   Community Health Advancement Initiative
CLOA   Certificate of Land Ownership Award
EU   European Union
FAO   Food and Agriculture Organization of the United Nations
FiBL   Research Institute of Organic Agriculture
FLO   Fairtrade International (formerly called Fairtrade Labeling Organiza-
         tions International)
GM   Genetically modified
ICAC   International Cotton Advisory Committee
IFAD   International Fund for Agricultural Development
IFOAM   International Federation of Organic Agriculture Movements
IMO   Institute for Marketecology
NCS   Nature Care Society
NGO   Non-governmental organization
NOFTA   Negros Organic Fair Trade Association
NOP   National Organic Program
PLA   Plantation Labor Act
Rs.   Indian Rupees
SFS   Surin Farmers Support
SIF   Social Investment Fund
STARFA   Santa Rita Farmers Multi-purpose Cooperative (formerly called Santa
         Rita Farm Workers Association)
USDA   the United States Department of Agriculture
WFTO   World Fair Trade Organization

1 Introduction

1.1 Two consumer movements surrounding agricultural products

The profusion of certification labels on processed food and beverage products, such as coffee and tea, from the global South perplexes consumers. There are primarily two categories of certification. The first provides consumers with an assurance that through their purchase of the labeled products they are helping disadvantaged Southern producers. The second provides an assurance that the products are environmentally friendly. Within the first category, the Fairtrade certification, issued by Fairtrade International (FLO), is the best known.[1] Organic certification is predominant within the second category. These two forms of certification have different origins and histories. Whereas Fairtrade certification is derived from the Fair Trade movement (see, e.g., Nicholls and Opal, 2005), organic certification is derived from the organic movement (see, e.g., Guthman, 2004a). However, by their designations appearing together as labels on the packaging of a single product, these separate movements have apparently been merged. According to a monitoring report released by FLO based on data collected in 2013 (Fairtrade International [FLO], 2014a, p. 62), 74 percent of all Fairtrade-certified producer organizations (including hired labor organizations) held at least one additional form of certification. Moreover, 51 percent of the Fairtrade-certified organizations held organic certification. Furthermore, Ecocert, which is a French inspection and certifying body, has introduced a single new integrated certification system that combines the Fair Trade and organic initiatives (French Fair Trade Platform, 2015, p. 16). An association of French companies, Bio Partenaire, has even created an "organic fair trade" label for its member companies engaged in organic and Fair Trade activities (French Fair Trade Platform, 2015, pp. 26–27).

Although the convergence of the two movements has been led by markets located in the global North, this has necessarily involved producers in the global South (Weber, 2007, pp. 113–114). As discussed in a subsequent section, whereas Fair Trade and organic agriculture differ when considered as movements, they share some characteristics in common when considered as certification systems. The similarities and differences between the two initiatives cause confusion not only among consumers in the North, but also among producers in the South.

This book is primarily concerned with the increasing interactions of these two global initiatives on the side of Southern producers.

At the onset, we would like to clarify certain key terms. In this book we use "movement" to express the original concept. This needs to be clearly differentiated from activities related to obtaining and maintaining "certification," which is a marketing tool of the movement. However, in practice there are situations in which outsiders cannot distinguish movement-related activities from certification-related ones. In such situations we use the term "initiative," which denotes a more comprehensive concept than either movement or certification. In other words, the Fair Trade initiative or "Fair Trade" includes all associated ideas and activities that pertain to the Fair Trade movement and certification. Similarly, the organic initiative includes both the organic movement and certification. Spelled as one word and capitalized, "Fairtrade" refers to the specific certification system implemented by FLO, that is, "a non-profit membership organization which defines fair production and trade standards, establishes auditing procedures and promotes the sale of labeled products" (Raynolds and Bennett, 2015, p. 6). Given the existence of similar fair and ethical trade programs, this book distinguishes between "Fairtrade" and "Fair Trade," which is a general term encompassing other similar programs.

Fair Trade, as defined in *A Charter of Fair Trade Principles*, which has been adopted by the World Fair Trade Organization (WFTO) and Fairtrade Labelling Organizations International (now known as Fairtrade International, FLO), is designed to "[contribute] to sustainable development by offering better trading conditions to, and securing the rights of, marginalized producers and workers – especially in the South" (WFTO and FLO, 2009, p. 4). Thus, the primary objective of Fair Trade certification systems is to help "marginalized producers and workers . . . in the South." Fairtrade certification is expected to provide certified producer groups with benefits such as higher prices for their products compared to free trade prices, minimum guaranteed prices, and price premiums that producer groups can take advantage of to improve their communal infrastructure and services (FLO, 2014b). In order to acquire Fairtrade certification and enjoy these benefits, participating producers are required to comply with regulations pertaining to "sustainable development." The phrase "sustainable development" connotes environmental, social, and economic sustainability (e.g., Hisas and Penaflor, 2006, p. 27). In addition to the components of social and economic sustainability embodied within cooperatives that are democratically organized and managed, as well as the ban on child labor, the same charter clearly refers to the environmental sustainability component as follows:

> All parties to Fair Trade relationships collaborate on continual improvement on the environmental impact of production and trade through efficient use of raw materials from sustainable sources, reducing use of energy from nonrenewable sources, and improving waste management. Adoption of organic production processes in agriculture (over time and subject to local conditions) is encouraged.
>
> (WFTO and FLO, 2009, p. 10)

It is highly feasible that producers who adopt organic production practices under a Fair Trade certification could also be certified as organic. Thus, a Fair Trade certification may contribute to the promotion of organic agriculture.

However, the organic movement is based on the "technical issue of production" (Browne, Harris, Hofny-Collins, Pasiecznik, and Wallace, 2000, p. 82). The Food and Agriculture Organization (FAO) of the United Nations (FAO, 2014) defines organic agriculture as follows:

> Organic agriculture is a holistic production management system which promotes and enhances agro-ecosystem health, including biodiversity, biological cycles, and soil biological activity. . . . This is accomplished by using, where possible, agronomic, biological, and mechanical methods, as opposed to using synthetic materials, to fulfill any specific function within the system.

Organic agriculture is evidently dedicated to the conservation of natural resources. However, consumers purchase organic food products for their personal health benefits rather than to contribute to environmental conservation (Cottingham and Winkler, 2007, pp. 32–33). Whereas consumers choose Fair Trade-certified products for altruistic reasons, they choose organic-certified products for reasons of self-interest. This may be one reason why the organic market is considerably larger than the Fair Trade market (Sahota, 2007, p. 25).

The global expansion of the organic movement also opened up opportunities for Southern agricultural producers to find new markets in the North for the products they cultivated (Raynolds, 2004). Many small and marginal farmers in the South traditionally follow organic cultivation practices that depend on locally available manures; conventional smallholders whose yields are comparatively low can significantly increase them by transitioning to organic production (Bacon, Méndez, and Fox, 2008, p. 353). The global organic movement offers small farmers in the South an opportunity to convert their prevailing disadvantages, such as small-scale production and the use of unsophisticated technologies, into positive sales points. Although organic certification does not require any social justice-related regulations, the International Federation of Organic Agriculture Movements (IFOAM) highlights "fairness" as one of the principles of organic agriculture:

> This principle emphasizes that those involved in Organic Agriculture should conduct human relationships in a manner that ensures fairness at all levels and to all parties – farmers, workers, processors, distributors, traders and consumers. Organic Agriculture should provide everyone involved with a good quality of life, and contribute to food sovereignty and reduction of poverty.
>
> (IFOAM, 2014)

The linking together of organic and Fair Trade certification systems appears to be logical. In practice, a few organic labels such as "Ecocert" and "Natureland" have adopted the Fair Trade component.[2] As in the case of Fair Trade certification systems, nations in the global North have led the organic certification process (Mutersbaugh and Klooster, 2011, p. 164).

Although Fair Trade–organic double certification has been a popular marketing practice (Allen and Malin, 2008; Bowes and Croft, 2007), the meaning of double certification has not been sufficiently explored. Most producer-focused studies have attended to only one of these forms of certification and have not distinguished one from the other (e.g., Getz and Shreck, 2006; Jafee, 2007; Thavat, 2011). An article by Browne et al. (2000) is noteworthy for introducing the discussion on the possible convergence of the two movements. Subsequently, some studies have sought to compare different certification systems (Giovannucci and Ponte, 2005; Kolk, 2013; Muradian and Pelupessy, 2005; Neilson et al., 2011). Parvathi and Waibel (2016) are exceptional in their examination of whether Fairtrade certification can bring additional benefits to organic-certified farmers, using panel data collected from black pepper cultivators in Kerala, India. Based on their comparison of single-certified organic farmers with double-certified farmers, they conclude that Fairtrade certification does not increase the incomes of organic farmers. The question this raises is this: Why did some farmers choose only organic certification, while others chose both kinds of certification? An empirical exploration of the different aspects of double certification would thus be pertinent.

1.2 The objectives and scope of the research

In this book we raise three questions concerning Fair Trade–organic double certification. First, under what conditions can producers obtain double certification? Second, do some farmers prefer single certification – Fair Trade or organic – to double certification? Third, how do these two forms of certification interact with each other?

We attempt to answer these three questions through five case studies from Asia. Many countries within Asia belong to the global South. Small farmers within the region, like those in Latin America and Africa, are seeking to benefit from the Fair Trade movement. However, traders and consumers in the North have little awareness of the realities of Fair Trade impacting Asian farmers and workers because the majority of studies on Fair Trade conducted by American and European scholars have tended to focus on Latin American and African producers. As we are Asian social scientists, our underlying aim in conducting these Asian case studies was to explore how Fair Trade has permeated Asian agriculture. This was our motive when we planned these case studies in Asia. Although organic agriculture was not the primary topic during the planning stage of the research, it appeared to be closely associated with Fair Trade in all of the case studies. The case studies subsequently revealed a variety of contradictions or tensions around Fair Trade and organic agriculture.

This book focuses on the convergence of two different certification schemes for Fair Trade and organic agriculture. As previously discussed, there have been attempts to incorporate both of these within a single certification label, such as

"Fair Trade by Ecocert" and "Natureland Fair" (French Fair Trade Platform, 2015, pp. 16, 64). However, because of the small scale of these efforts, it is difficult to find cases for observation, especially within Asia. We believe that the findings of our study, which focuses on the convergence of these two certification systems, will be conducive to the development of such a single merged certification system.

Before presenting the case studies, it would be salient to provide an overview of these respective initiatives within Asian agriculture and the presence of Asia within the studies on both Fair Trade and organic agriculture.

1.3 The Fair Trade initiative in Asian agriculture

Monitoring reports issued by FLO (2013, 2014a) provide an outline of the development of Fair Trade in Asia. The first characteristic is that "Fairtrade has historically been smaller in scale and slower to grow in Asia than elsewhere" (FLO, 2013, p. 7). As Table 1.1 shows, only 17 percent of all Fairtrade farmers and workers live in Asia (FLO, 2014a, p. 18).[3] However, in recent years Asia has shown the highest growth rate in terms of the number of Fairtrade farmers and workers (FLO, 2013, p. 18). From 2012 to 2013 the number of Fairtrade-certified producer organizations increased by 12 percent in Asia, compared with a 5 percent increase in other regions (FLO, 2014a, p. 37). At the end of 2013 there were 182 Fairtrade-certified producer organizations distributed in 18 countries in Asia: of these, 115 were small farmer organizations, 19 were contract production organizations, and 48 were hired labor organizations (FLO, 2014a, p. 144).[4] Thus, Asia's Fairtrade production is characterized by a large portion of hired labor organizations, represented by tea plantations. Asia leads in the area of plantation-based Fair Trade. Specifically, this region accounts for 80 percent of the world's Fairtrade tea plantation workers (FLO, 2014a, p. 108). Thus, whereas Asia supplies 49 percent of all Fairtrade workers in the world, it accounts for just 12 percent of all Fairtrade farmers (Table 1.1).

Table 1.1 The global distribution of Fairtrade farmers and workers, 2013 (unit: the number of registered persons)

Region	Fairtrade farmers	%	Workers on Fairtrade-certified plantations	%	Total	%
Latin America and the Caribbean	309,500	24	13,900	7	323,400	21
Africa and the Middle East	838,500	64	93,600	44	932,100	62
Asia and Oceania	157,500	12	103,400	49	260,900	17
World	1,305,500	100	210,900	100	1,516,400	100

Source: FLO (2014a, p. 18)

India is Asia's largest supplier of Fairtrade producers within both categories of organizations, comprising small farmers and hired labor respectively. A total of 80 (48 percent) out of 182 certified Asian organizations are located in India (Table 1.2). India ranks fifth globally for the category of small farmer organizations, followed by Indonesia, Thailand, and China in descending order (Table 1.2; FLO, 2014a, p. 20). For the category of hired labor organizations, India is the largest supplier of Fairtrade workers globally, followed by Kenya and Sri Lanka (FLO, 2014a, p. 20). A monitoring report released by FLO (2013) highlights the particular "importance of Indian tea and seed cotton, coffee from Indonesia and Papua New Guinea, and the rapid growth in importance of sugar from Fiji" (p. 112). Moreover, India is now expected to be an important supplier not only of Fairtrade producers, but also of Fairtrade consumers. At the end of 2013 a Fairtrade marketing organization was also established in India to promote and market Fairtrade products to domestic consumers. Unlike Northern marketing organizations that deal with Fairtrade products from around the world, this Indian marketing organization focuses on markets that are exclusively for Indian Fairtrade products (Fairtrade India, 2013).

The large presence of tea plantations, which is an Asian characteristic, is also evident in Fairtrade premium distribution by product, which provides an indication of Fairtrade product sales (Table 1.3). In Asia, 21 percent of this premium comes from tea, whereas tea constitutes only 5 percent of the global Fairtrade premium. Because coffee has the largest share of the Fairtrade market, its share of

Table 1.2 Fairtrade producer organizations in Asia, 2013

Country	Small producer organization	Contract production	Hired labor organization	All types
Afghanistan	2			2
China	14			14
Fiji	3			3
India	32	18	30	80
Indonesia	17			17
Iran	1			1
Kyrgyzstan	1			1
Lao PDR	1			1
Nepal			1	1
Pakistan		1	5	6
Papua New Guinea	5			5
The Philippines	4			4
Samoa	1			1
Sri Lanka	6			18
Thailand	15			15
Timor-Leste	1			1
Uzbekistan	3			3
Vietnam	9			9
Total	115	19	48	182

Source: FLO (2014a, p. 147)

Table 1.3 Fairtrade premium distribution by product, 2012–2013

	Global total		Asia and Oceania*	
	Amount (Euros)	Share (%)	Estimated amount (Euros)	Estimated share (%)
Coffee	43,960,700	46.2	4,725,000	45.0
Bananas	17,018,900	17.9	0	0
Cocoa	9,828,500	10.3	105,000	1.0
Cane sugar	9,790,300	10.3	2,415,000	23.0
Flowers and plants	5,096,000	5.4	N/A	N/A
Tea	4,552,100	4.8	2,205,000	21.0
Fresh fruits	1,395,800	1.5	105,000	1.0
Wine grapes	873,600	0.9	0	0
Seed cotton	644,000	0.7	525,000	5.0
Herbs, herbal teas, and spices	450,300	0.5	N/A	N/A
Honey	319,900	0.3	N/A	N/A
Rice	296,200	0.3	315,000	3.0
Quinoa	228,300	0.2	0	0
Nuts	202,700	0.2	N/A	N/A
Dried fruits	152,600	0.2	N/A	N/A
Fruit juices	137,800	0.1	105,000	1.0
Oilseeds and oleaginous fruits	92,800	0.1	N/A	N/A
Sports balls	87,000	0.1	105,000	1.0
Gold	64,100	0.1	0	0
Vegetables	49,300	0.1	0	0
Total	95,240,900	100.0	10,500,000	101.0

Source: FLO (2014a, pp. 66–67, 144–146)

Note: *Fairtrade International publishes only rounded figures for the total amount (10,500,000) and percentages of major products. Therefore, there are some errors in the calculated amounts. The products for which data were not available (denoted by N/A in the table) comprised approximately 1 percent of the total amount.

the premium is correspondingly the largest in Asia (45 percent), as well as globally (46 percent). The premium distribution also shows that other major crops in Asia are cane sugar (23 percent), seed cotton (5 percent), and rice (3 percent). These three crops each surpass the respective crop's global average in the share of premium distribution.

The FLO monitoring reports suggest that Asian producer organizations are not as successful at selling their products within Fairtrade markets located in the North as those belonging to other regions. In 2012 Asian producer organizations, representing 18 percent of Fairtrade farmers and workers globally, obtained just 11 percent of all Fairtrade premium revenues (FLO, 2013, p. 111). This trend continued in 2013 (FLO, 2014a, p. 70). This gap between the workforce and premium revenues suggests that after obtaining certification many Asian producer cooperatives and plantations may find it difficult to sell a significant percentage of their outputs as Fairtrade products. Moreover, this

Table 1.4 The average Fairtrade premium per farmer or worker by region, 2012–2013

Region	Amount (Euros)
Central America	198
The Caribbean	220
South America	229
Northern Africa and the Middle East	43
Western Africa	48
Eastern Africa	20
Central Africa	14
Southern Africa	154
Eastern Asia	**149**
Southeastern Asia	**73**
Southern Asia	**19**
Central Asia	**15**
Oceania	149
World average	73

Source: FLO (2014a, p. 72)

gap is even wider for the category of hired labor organizations. The recent FLO report notes: "While workers in plantations . . . in Asia and Oceania account for 49 percent of all workers in the Fairtrade system, only 12 percent of the global Fairtrade premium flows to these organizations" in the region (FLO, 2014a, p. 144). As Table 1.4 indicates, the premium accrued by Asian farmers and workers is much smaller than that accrued by farmers and workers in other regions. While the number of farmers and workers in Southern Asia is large, the amount of premium they receive on an individual basis is very small. Given the limited size of the Fairtrade market, a crucial problem facing Asian Fairtrade producers appears to be one of how to successfully compete against Fairtrade producers from other regions.

Although limited information is available on other Fair Trade activities, Asia's contribution in these activities seems to be negligible. The *International Guide to Fair Trade Labels* reports that the French "Fair Trade by Ecocert" program has certified 217 bodies globally, out of which only 7 organizations (3 percent) are located in Asia (French Fair Trade Platform, 2015, p. 17). Similarly, the "Fair for Life" label, created by a Switzerland-based certification body, Institute for Marketecology (IMO), to combine organic farming, corporate social responsibility, and Fair Trade, has been given to just one organization in Asia (1 percent), compared with a total of 100 organizations across the globe (French Fair Trade Platform, 2015, p. 29).

1.4 The organic initiative in Asian agriculture

Organic agriculture still constitutes a minor component of agricultural production in Asia. According to the most comprehensive report issued by the Research Institute of Organic Agriculture (FiBL) and the IFOAM (Willer and Kilcher, 2012, p. 38), only 7.5 percent of the global area of land used for certified organic agriculture was located in Asia in 2010. Whereas the global ratio of organic agricultural land to total agricultural land was 0.9 percent, the ratio specifically in Asia was 0.2 percent (Table 1.5).

Asia is evidently divided into two kinds of countries: those that are organic producers and those that are organic consumers. Countries engaged in large-scale cultivation of organic crops are China, India, Thailand, Indonesia, the Philippines, and Vietnam, where organic products are mainly grown for export (Orboi, 2013, p. 208). China has the largest acreage of land dedicated to certified organic agriculture, followed by India, Kazakhstan, the Philippines, and Indonesia (Table 1.6). Important organic agricultural exports include fruits,

Table 1.5 Region-wise areas of organically cultivated land (certified and in-conversion), 2010

Region	Organically cultivated land (hectares)	Distribution by region (%)	Ratio to total agricultural land (%)
Africa	1,075,829	2.9	0.1
Asia	**2,778,291**	**7.5**	**0.2**
Europe	10,002,087	27.0	2.1
Latin America	8,389,459	22.7	1.4
Northern America	2,652,624	7.2	0.7
Oceania	12,144,984	32.8	2.9
Global total	37,041,004*	100.0	0.9

Source: Willer and Kilcher (2012, pp. 38, 42)

Note: *Includes corrections.

Table 1.6 Country-wise areas of organically cultivated land within Asia (certified and in-conversion), 2010

Country	Organically cultivated land (hectares)	Ratio to total agricultural land (%)
China	1,390,000	0.27
India	780,000	0.43
Kazakhstan (2009)	133,562	0.06
The Philippines	79,992	0.67
Indonesia	71,208	0.13
Sri Lanka	43,664	1.67
Saudi Arabia	42,376	0.02
Thailand	34,079	0.17
Timor-Leste	24,750	6.60
Pakistan	22,103	0.08
Azerbaijan	21,347	0.45
Syria	19,987	0.14
Vietnam	19,272	0.19
Republic of Korea	15,518	0.84
Kyrgyzstan	15,040	0.14
Others	65,394	-
Total	2,778,292	0.20

Source: Willer and Kilcher (2012, p. 195)

vegetables, herbs, spices, rice, and tea (Willer and Kilcher, 2012, p. 125). By contrast, the demand for organic products is concentrated in affluent countries, including Japan, South Korea, Taiwan, Hong Kong, and Singapore; despite the high demand for organic products, their production is relatively limited in these large-scale consumer countries (Willer and Kilcher, 2012, p. 125). Japan and Taiwan feature in the "others" depicted in Table 1.6.

There has been remarkable progress in organic certification in Asia's producer countries (see Figure 1.1). Of the total number of 549 organic certification bodies globally, 179 were located in Asia (as of 2011). A rapid increase in this number has been observed within some Asian countries, particularly India (Orjavik, 2012, p. 137). Of the 179 Asian organic certification bodies, 61 are located in Japan, 33 in South Korea, 28 in China, and 22 in India (Orjavik, 2012, p. 137). Local certification bodies in other Asian countries are relatively small and find it difficult to compete with international certification bodies (Wai, 2012, p. 174). Only the accreditation systems of India and Japan have been recognized by the European Union (EU) and the United States Department of Agriculture (USDA) (Wai, 2012, p. 174). Consequently, "[m]ost exports elsewhere are certified by international certification bodies working in the regions accredited by international and EU-based accreditation bodies or directly by the USDA" (Wai, 2012, p. 174). India seems to be the leading Asian country in terms of Fair Trade–organic double certification.

The International Fund for Agricultural Development (IFAD) has characterized organic agriculture in Asia as being "very much a smallholder-oriented endeavor" (IFAD, 2005, p. 2). Although, as evidenced by statistical data, the number of certified organic farms producing for a premium-price market is small, many non-certified organic farms appear to conduct organic farming practices for

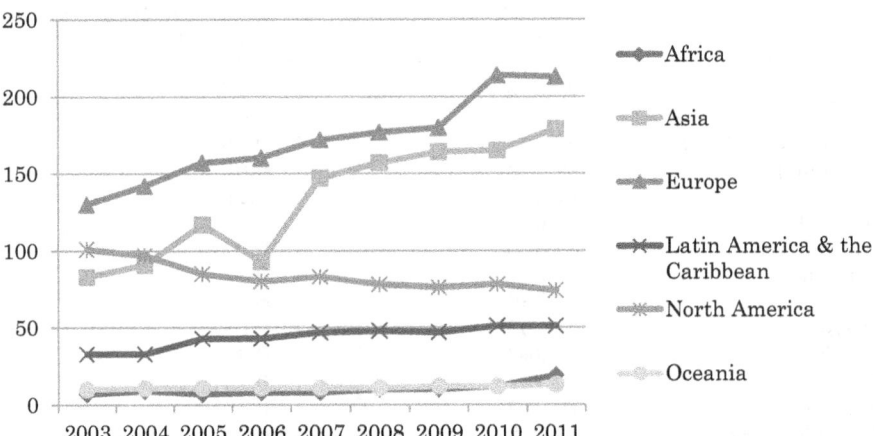

Figure 1.1 Region-wise development of organic certification bodies, 2003–2011
Source: Willer and Kilcher (2012, p. 139)

their own households and for local markets. While it is difficult to estimate the extent of informal organic agriculture in terms of farming populations and land areas, this may considerably exceed that of certified organic agriculture (Halberg and Muller, 2013, p. 7). Small farmers in Asia "straddle the two categories" of certified and non-certified organic farming, "or sometimes slide between them from year to year" (IFAD, 2005, p. 2).

In many Asian countries organic agriculture has taken root and expanded as a result of grassroots efforts. Non-governmental organizations (NGOs) and farmers' groups have played a vital role in these efforts through their dissemination of information and knowledge regarding organic cultivation, as well as through their facilitation of market access (IFAD, 2005, p. 6). As IFAD noted (2005, p. 4), "[o]nly later, as the potential benefits of organic agriculture for small farmers became more widely appreciated, did the government[s] participate more actively in its development."[5]

1.5 Asia in Fair Trade studies

Given the small ratio of Asian Fair Trade producers, studies on Fair Trade have correspondingly paid scant attention to Asia. A recent study by Parvathi and Waibel (2016) focused on double-certified black pepper farmers in India (see Section 1.1). In addition, studies by Neilson and Pritchard (2009) and by Besky (2014) also focused on India as the largest supplier of Fairtrade-certified producer organizations, although Fair Trade was not necessarily central to these studies. Neilson and Pritchard (2009) showed how the coffee and tea production sectors within South India are involved in global value chains, some of which draw on Fairtrade certification. Their study suggested the importance of analyzing the impact of Fair Trade on producers within these chains from the wider perspective of place-specific institutional environments. Besky (2014) ethnographically explored the lives of tea plantation workers in Darjeeling, examining how notions of fairness, value, and justice have shifted with the rise of Fair Trade practices and postcolonial separatist politics. Her conclusion is that Fair Trade on tea plantations has amplified, rather than minimized, the hierarchical nature of the traditional plantation society. While such studies provide us with some information on Indian Fair Trade producers, especially in relation to certified tea estates, the realities of Fair Trade within Asia more broadly remain to be explored.

1.6 Asia in organic agricultural studies

Although organic agricultural studies have paid more attention to Asia compared with studies on Fair Trade, most of these studies have adopted agronomic and technological approaches to organic agriculture in Asia (e.g., Krasachat, 2012; Nayak et al., 2012). A few social scientists who have adopted global perspectives have included parts of Asia in their investigations.

For example, Knight and Newman (2013) used cross-national data, including data from Asian countries, to empirically assess the environmental consequences of organic agriculture. They found that the intensity of agricultural export is positively associated with the area of land under organic agriculture. Barrett, Browne, Harris, and Cadoret (2002) identified the obstacles to certification that confront organic producers in the developing world, including Asia. They concluded that cost-effective certification is essential for the expansion of organic exports from developing countries. Insights gained from these cross-national studies are significant within the Asian context. However, data on the performance of organic agriculture in Asia are limited. Thus, Bennett and Franzel (2013) were compelled to confine their synthetic examination to cases of African and Latin American farmers who converted from conventional to organic farming.

With the exception of Sano and Prabhakar (2010), who proposed national organic promotion policies from a macroscopic viewpoint, Asia-focused studies have tended to engage with specific countries and products. Carpenter (2003) and Eyhorn (2007), both focusing on organic farming practices, examined the conversion from conventional to organic farming in the cases of rice cultivation in the Philippines and cotton cultivation in India. Both studies found that organic farming benefits resource-poor farmers. Conversely, other studies have offered different kinds of criticism of certified organic agriculture. For example, Thiers (2005), whose study was conducted in the context of the authoritarian political economy of rural China, argued that contradictions between ecological and market modes of rationality that are inherent in organic certification and marketing systems may not promote ecological agriculture. A study by Eernstman and Wals (2009), conducted on a tribal society in India, demonstrated that the introduction of universal principles and standards for certification "may actually be counterproductive in the development of contextually appropriate and workable sustainable land management systems" (p. 375). Moreover, Thavat (2011) argued that promoting organic rice cultivation in Cambodia, against agrarian changes such as the significant increase in off-farm employment, helps only those farmers who are fortunate to have adequate labor and limited or seasonal off-farm work opportunities. A review of these case studies provides us with insights on some of the emerging contradictions in the transition from traditional or natural farming to certified organic farming.

1.7 Contradictions from different perspectives

Whereas preceding studies have revealed some contradictions embedded in each of the Fair Trade and organic agriculture initiatives, further contradictions or tensions can be observed when organic agriculture is combined with the Fair Trade initiative. To reveal the realities of double certification, this book draws on contradictions faced by current and prospective beneficiaries of the two movements – small farmers and plantation workers in Asia. Some contradictions are apparent in relation to the different natures of the two movements, whereas others may be associated with the disparity between the original objectives of a movement and the commercialized tool of the movement, that is, certification.

As the conceptual mapping provided in Figure 1.2 shows, theoretically, contradictions could emerge in different places (A~K) through the combination of the Fair Trade and organic initiatives. Most critiques of Fair Trade (e.g., Besky, 2014; Getz and Shreck, 2006) have highlighted contradictions that exist between the movement and certification (at location F in Figure 1.2). These critical studies have focused on a movement's transformation into commercial certification, revealing how the original concept that anchored the movement is lost in the transformation process. Guthman (2004a) referred to this transformative process in relation to organic agriculture (at location H in Figure 1.2) as "conventionalization." Critiques of organic certification systems in Asia (Eernstman and Wals, 2009; Thavat, 2011; Thiers, 2005), discussed in Section 1.5 of this chapter, can also be conceptually mapped within location H. However, contradictions may be observed within a movement (locations A and B in Figure 1.2) as well as within a

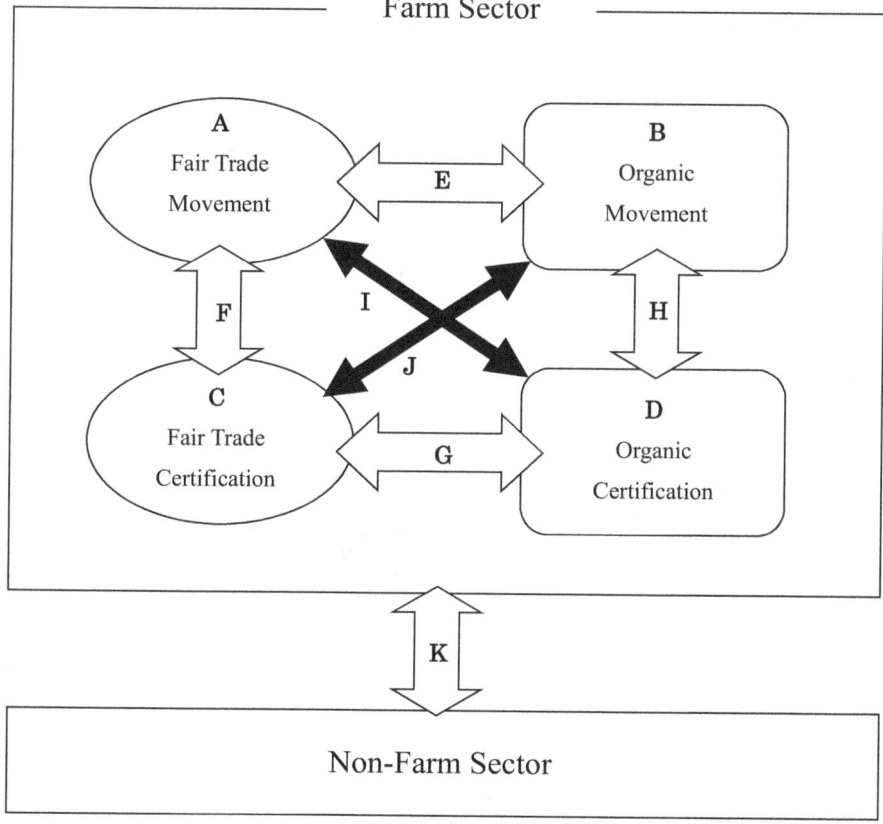

Figure 1.2 A conceptual mapping of contradictions or tensions arising from the Fair Trade and organic initiatives

certification system (locations C and D in Figure 1.2). Furthermore, considering the growing significance of non-farm income sources (e.g., Rigg, 2006), the promotion of intensive farming through Fair Trade and organic initiatives may entail tensions relating to the compatibility of these initiatives with non-farm activities (location K in Figure 1.2). Contradictions surrounding the two initiatives can thus occur at various locations.

In this book we attempt to identify different types of contradictions or tensions through five case studies in Asia. Three of the case studies are from India, Asia's largest supplier of Fairtrade products; one case study is from Thailand; and the last one is from the Philippines. One of the Indian cases relates to the plantation sector, defined by FLO as a hired labor organization, whereas the other two cases relate to the small farmer sector. The following export crops feature in these studies: coffee, tea, cotton, rice, and sugarcane. Makita, the first author, conducted the case studies in India and the Philippines, and Tsuruta, the second author, conducted the case study in Thailand. Figure 1.3 depicts the locations of the case studies. Shedding light on different types of contradictions, these case studies reveal significant issues that are obscured by the Northern consumers-led Fair Trade and organic movements.

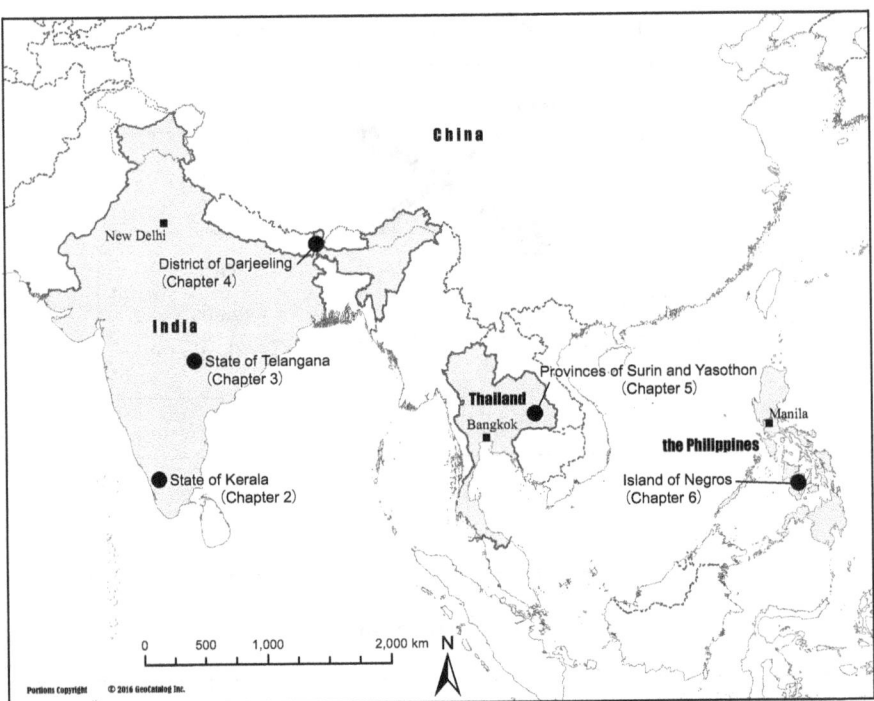

Figure 1.3 Locations of the case studies

1.8 Outline of the book

In Chapter 2, we begin with an examination of a basic contradiction that prevails between Fair Trade and organic certification systems at the production level. Fair Trade certification requires farmers to focus on specific products, whereas organic agriculture targets production units. This chapter explores how small-scale producers in the South have contended with this discrepancy through the observation of a South Indian farmer group's attempts to obtain Fairtrade and organic certification. This case study reveals variations in the responses of stakeholders, pursuing different livelihood strategies, who reacted to the two forms of certification differently. The contradiction that emerged in this case can be interpreted in terms of two contradictions that exist between the organic movement and Fairtrade certification (depicted at location J in Figure 1.2) and between the organic and Fairtrade certification systems (location G in Figure 1.2). An important finding from this case study is that each farmer's perception of the contradiction depends on his or her livelihood strategy. For Southern producers, joint adoption of the two initiatives does not always provide better opportunities than the separate adoption of either one of them does.

Chapter 3 explores another convergence of Fair Trade and organic initiatives in the specific context of genetically modified (GM) cotton cultivation in India, which is the world's second-largest cotton-producing country. In the case of small cotton producers in central India, contrary to the ethical precepts of the two initiatives, their convergence has contributed to the spread of GM seeds. This case study reveals that a more relaxed attitude toward GM crops is apparent within the Fair Trade initiative during the period of conversion from conventional to organic farming. The Fair Trade initiative lacks any compensatory scheme for redressing the decrease in incomes that producers have to endure during this conversion period. Nevertheless, the initiative has remained faithful to its original mission of assisting small farmers. The entailed Fair Trade dilemma, arising from a focus on helping poor farmers and promoting organic farming, may have indirectly contributed to the tendency of Fair Trade producers to engage in conventional farming using GM seeds. Although the Fair Trade initiative itself addresses the promotion of organic agriculture, this case study, in which a third lens of GM seeds was applied, reveals a clear difference between the two movements. While they appear to be closely linked, when they converge they may in fact counteract each other. This contradiction, revealed in Chapter 3, can be mapped at location E in Figure 1.2.

Chapter 4 attends to a contradiction within India's tea plantation sector. This pertains to Fairtrade certification introduced for plantations (location C in Figure 1.2) and is particularly evident in relations between a plantation's managers and workers. This case study analyzes the impacts of Fairtrade certification, focusing on its interactions with traditional patron–client relations between the management and workers within India's tea plantations. The case study reveals that Fair Trade is not known to workers, and, further, that this invisibility of

Fair Trade has generally reinforced existing patron–client relations through Fairtrade premiums. Organic certification has also contributed to the reinforcement of existing power relations. However, the case study shows that when a Fairtrade premium is invested in a community development project led by an independent third-party organization, the invisibility of Fair Trade inversely hides the management's patronage. This case study suggests that the identified contradiction not only may be a constraint for plantation workers, but also may open up a new opportunity for them.

In Chapter 5 we present a case wherein an anticipated contradiction did not occur in some situations. Many authors have argued that the introduction of certification has transformed the Fair Trade and organic movements from ethical to business-oriented initiatives (e.g., Raynolds, Murray, and Taylor, 2004). This type of transformation can be mapped within the F or H locations shown in Figure 1.2. However, our case study in Thailand shows that while some groups of organic farmers in northeastern Thailand were able to retain the original visions of the movements and to put them into practice, others, like many of the producers identified in previous studies, took advantage of certifications solely to benefit their businesses. The case study illustrates how two rice producers' groups have adapted the organic and Fair Trade movements, founded in the West, to the local socioeconomic context by comparing the two groups with a typical business-oriented group. Key members of the two groups were able to retain the ethical foundations of these initiatives, even with the introduction of commercialized certification schemes. It turned out that Fair Trade and organic movements in the region had evolved out of a preexisting rural campaign initiated by Buddhist monks that sought to foster a fair and sustainable society. In other words, the movements were merged with ideas and practices emerging from within the society and derived from a long-established Buddhist tradition.

The above-mentioned chapters focus on contradictions existing within the agricultural sector. However, the majority of farmers who participate in the Fair Trade and organic initiatives also attempt to diversify their livelihoods into non-farm activities in parallel with farm activities. Consequently, Chapter 6 focuses on the tensions between certification-supported farming and other non-farm livelihood activities. Conceptually, this type of tension can be mapped in location K in Figure 1.2. Through the observation of a sugarcane producer cooperative comprising land reform beneficiaries in the Philippines, this chapter explores how households that are members of the certified cooperative achieve compatibility between their certification-supported farming and the diversification of individual livelihood activities. In this case, livelihood diversification at the household level progressed both away from farming and into farming. Underlying the long-term use of agricultural certification were compelling reasons for continuing both certification-supported farming and diversified livelihood activities. Moreover, mechanisms existed whereby cooperative members, regardless of their level of diversification, were able to achieve compatibility between both sets of activities. This case study suggests an effective way of taking advantage of agricultural certification for small farmers.

In the final chapter, we summarize the different kinds of contradictions or tensions that can arise with the introduction of Fair Trade and organic initiatives within small farmers' production activities in the global South. We subsequently reexamine double certification through a cross-case analysis and address the three opening questions that we raised at the beginning of this chapter. Last, from a wider perspective we discuss the possible role that the two initiatives, at their confluence, can play in rural development in the global South. Fair Trade–organic double certification may potentially serve as an intermediary institution, linking the two key objectives of rural development, namely, poverty reduction and environmental conservation.

Notes

1 Fairtrade International was previously known as Fairtrade Labeling Organizations International (FLO). Even after the organization was renamed the abbreviation FLO was retained.
2 In 2010 the Germany-based Natureland Association for Organic Agriculture launched the label "Natureland Fair" as an additional option for companies already using the organic label "Natureland" (French Fair Trade Platform, 2015, p. 64).
3 In the reports issued by FLO, Asia includes the Pacific islands located in Oceania.
4 "Contract production organization" is a special category that applies to rice and cotton producers in India, cotton and dried fruit producers in Pakistan, and cocoa producers in Oceania (FLO, 2014a, p. 42).
5 China is a notable exception. Chinese local and provincial governments invested in a number of successful export-oriented organic agriculture enterprises during the early stages of the organic initiative (IFAD, 2005, p. 4).

2 Fair Trade and organic agriculture

A case from South India

This chapter sheds light on a fundamental difference between Fair Trade and organic certification systems – the difference in *what* is certified. Producers' reactions to this difference can be observed only in the process of pursuing both certifications. The observation of already-double-certified producer groups – a successful outcome of the process – may not reveal some contradictions confronting producer groups that once pursued double certification and failed as a result. An NGO's project implemented for small farmers in South India provides a suitable case for observing the impact of the two certifications' difference on producers.

2.1 A crucial difference between Fair Trade and organic certification systems

It is worthy to review what is actually certified in the respective certifications. A Fair Trade label is provided for a specific product in the agricultural sector.[1] This is mainly because Fair Trade labeling has been developed to offer consumers an important degree of reassurance when buying such products (Nicholls and Opal, 2005, p. 11). Even if the product certification is criticized for leading to monoculture, the potential range of Fair Trade products is in reality restricted, owing to the limited acceptance of many basic Southern products in Northern diets (Wilkinson and Mascarenhas, 2007). If a Fairtrade-certified cooperative with coffee wants to sell another product in addition to coffee, they need to meet the different standard specifically for the additional product and pay the additional fee for the product (FLO-CERT, 2015).

In contrast, organic certification assures the organic nature of production activities by a certified farmer or farmers' group. In other words, "all products from all environmentally friendly production units will be automatically labelled" (Dankers, 2003, p. 23). This accreditation system is based on the principle of organic farming, that is, "a holistic production management system which promotes and enhances agro-ecosystem health" (FAO, 2014). Therefore, certified farmers and farmers' groups can sell any crop as organic from their "environmentally friendly production units" after they obtain certification.

This fundamental difference in the nature of certification may draw different reactions from Southern farmers to the two initiatives. Organic certification may help farmers to consider biodiversity and long-term soil fertility in their entire farmland, rather than to increase the production of a specific organic crop. When the same farmers attempt to break into a Fair Trade market, they need to select one suitable crop, which will inevitably require them to pay special attention to the selected crop. The confluence of Fair Trade and organic agriculture is likely to make Southern producers face a dilemma: entire production unit or specific crop. When pursuing double certification, how does each small farmer cope with this dilemma?

2.2 Outline of the case

2.2.1 An NGO's project in South India

One of us, Makita, observed a project led by an Indian NGO called Agriculture and Organic Farming Group India (AOFG). The project aimed to organize an association of small farmers who own fewer than five acres in highland Kerala in southern India and to convert them into organic farmers. In India only 30 percent of the total cultivable area is irrigated and treated with fertilizers; in the remaining 70 percent, which is mainly rain-fed, negligible amounts of fertilizer are used (Thapa and Tripathy, 2006, p. 93). With the sizable acreage under naturally organic cultivation, India has tremendous potential to emerge as a major supplier in the world's organic market (Thapa and Tripathy, 2006). The state of Kerala is one of the most advanced states in certified organic farming; IndoCert, a domestic certification body, has issued the largest number of certificates to Kerala farmers or farmers' groups (based on an interview in IndoCert, August 2008).

The goal of the small farmers' association was to export their certified crops to foreign organic markets. When the author first visited the project site in November 2008, it had been only seven months since the official establishment of the small farmers' association, while AOFG had spent more than two years motivating and mobilizing small and marginal farmers. A variety of activities were still in an experimental stage. About 10 to 15 farmers composed one group; 52 groups had already been formed. A majority of members were below the national poverty line. Member farmers cultivated a wide range of crops, including coffee, rubber, cocoa, pepper, ginger, nutmeg, tapioca, and bananas, in small plots of steep land.

AOFG had provided three categories of intervention: (a) the promotion of organic farming, (b) special assistance for a potential Fair Trade product (coffee), and (c) other activities for organizing farmers' groups.

For the purpose of obtaining a group organic certification, AOFG introduced a liquid bio-fertilizer to members. At the start of AOFG's project some members depended on chemical fertilizers, while some were traditionally organic farmers relying only on cow dung, litter, and crop residues. The introduced bio-fertilizer was effective not only as a substitute for chemicals but also as a supplement to traditional manures. It was expected to cure the diseases of some crops, as well as to increase productivity at a cost lower than that required for chemicals.[2] Out of many options, AOFG selected the ready-for-use liquid fertilizer, easiest to prepare at the farm level. In general, one reason why many farmers prefer chemical fertilizers to organic manures is that the former are soluble in water, are easily absorbed by plants, and therefore give instant results; in contrast, it takes a long time for organic manures applied with elements in insoluble forms to interact with microbes and show visible effects (Dhavse, 2004). The liquid bio-fertilizer could solve this problem. An appointed member of the association took charge of mixing compost, liquid manure, bio-manure powder, sugar, and water in a special refiner, which was also supplied through AOFG. The completely concentrated bio-fertilizer was then distributed to each member (Figure 2.1). Member farmers had only to dilute the liquid with water before applying it to crops within 24 hours.

In parallel with the introduction of the new bio-fertilizer, AOFG started to train internal inspectors. Provisional internal inspectors, chosen from member farmers, periodically visited members and monitored their use of fertilizers. AOFG also provided member farmers with extension services in organic farming, which could rarely be expected from government departments because of the limited numbers of field officers. Through AOFG member farmers were able to readily collect information relevant to organic farming: for instance, details of government subsidies for the purchase of organic fertilizers.

To become completely organic is a long process. Only after obtaining an organic certification are member farmers allowed to sell their products to organic

Figure 2.1 Concentrated bio-fertilizer taken out of the refiner for distribution (Kerala, India)

markets. AOFG considered it necessary to give member farmers some tangible benefit from being organic even during the transitional period. AOFG therefore decided to pursue a Fairtrade label for coffee from member farmers. Coffee, which is a major Fairtrade item, has traditionally been a cash crop in highland Kerala. To show a benefit from participation in the farmers' association, in November 2008 AOFG began to purchase fresh, ripe coffee berries from members at a higher-than-market price near their farms.[3] This helped farmers to save costs and time for post-harvest processing and transportation. AOFG established a large drying space and a primary processing unit in the project site and was planning to build a processing factory near Cochin (a major port city in southern India) for the production of final coffee products for export. AOFG also nurtured seedlings of a specific variety suitable for the selected processing method and distributed them to member farmers at a subsidized price.[4] Although AOFG purchased coffee from member farmers, it was impossible to export unless the processing factory was in operation and the farmers' association obtained organic or Fairtrade certification. AOFG actually sold collected coffee to a local non-organic market in the season studied.

To organize small farmers into groups and to reinforce organized groups, AOFG also provided other forms of technical and financial assistance for the farmers' association. Most importantly, AOFG helped each group to open a group savings account in a commercial bank. Through this group account members were expected to access bank loans for their own investment purposes in the future. This scheme was beneficial for many small farmers who did not individually have access to formal loans due to the lack of formal land titles. As of December 2008 most of the 52 groups had started their regular savings, and 17 groups had already opened bank accounts. In addition, AOFG financially supported the management of the farmers' association by paying regular remuneration to the committee members, who were selected from members.

2.2.2 A focus group of farmers and data collection

Makita's field study focused on one group consisting of thirteen member households who were neighbors in a village. Focusing on one group enabled her to observe their livelihood strategies closely. This group was chosen as a result of discussion with the AOFG fieldworkers. It met all three criteria for selection: (a) it was not among the most active groups, (b) it was not among the most inactive groups, and (c) it included members of a "scheduled tribe" unique to the area. According to Srivastava (2008), scheduled tribes are generally characterized by their rustic way of living, distinctive culture, geographic isolation, shyness of contact with the outside world, and economic and social vulnerability. The majority of tribal people are marginal farmers because in the long colonial and postcolonial eras they were deprived of their land by British and Indian settlers (Kjosavik and Shanmugaratnam, 2007). Eight of the thirteen families had only possession rights of their land, in place of formal titles. Three of the thirteen families had no electricity.

In November and December 2008, primary data were collected on the members' reactions to the interventions for organic and Fairtrade certifications through semi-structured interviews with all adults in the member families. Supplementary interviews were conducted with fieldworkers of AOFG, other member and non-member farmers, and local traders dealing with major cash crops. Analysis of the collected data aims to clarify how participant farmers incorporated the new interventions into their livelihood strategies. AOFG's interventions were used only to observe member farmers' reactions to them. To evaluate the effect of each intervention, such as the quality of the selected bio-fertilizer, was beyond the scope of this research.

2.3 A real confluence of Fair Trade and organic agriculture: Four types of farmers in the group

While all thirteen members were small farming families who cultivated between 0.5 and 4.0 acres of land, they can be classified into four categories: (a) five full-time subsistence farmers, including the president of the group; (b) a full-time commercial farmer; (c) three wealthier part-time farmers; and (d) four poorer part-time farmers. Table 2.1 summarizes major characteristics of each category.

Small family size was common to all the full-time subsistence farmers. Three of the five families were middle-aged or elderly couples, and their children were financially independent; another elderly couple lived with a mentally disabled son; and one single woman took care of a nephew in place of his father, who lived in another village. Because of the small numbers of dependents, these five families were able to somehow make a living by subsistence farming on their small plots of land.

One full-time farming family was overtly distinct from the above-mentioned subsistence farmers. The family head, called KJ, his wife, father, and mother lived

Table 2.1 Classification of the group member households (Kerala, India)

	Type A 5 full-time subsistence farmers	Type B One commercial farmer	Type C 3 wealthier part-time farmers	Type D 4 agricultural labor-cum-farmers
Largest income source	Farming	Farming	Non-farming	Hired labor
Number of dependents	Smaller	Smaller	Smaller	Larger
Encroached land	Not available	Available	Not available	Not available
Access to bank loans	Available: 1 Not available: 4	Available	Available	Not available
Cash liquidity	Higher	Lower	Lower	Highest

(13 member households in total)

in the same house, and both KJ and his father were members. Therefore, correctly speaking, this group was composed of 14 individual members. While this family officially owned 3 acres of farmland, they had customarily encroached on and used more than 20 acres of forestland for coffee cultivation. This unofficially occupied land had brought large profits to this family for the past 20 years.

The other seven families had another major income source in addition to farming on their own land. Three of them had stable employment. In one family the husband was permanently employed by a government organization, the Rubber Board, and his wife was a primary-school teacher. This family was maintaining farming by employing many casual laborers. The head of another family worked for a private bank, and his resident son took charge of farming. In the third family, the husband worked as a load carrier in the nearest town, and his wife ran a cafeteria in the village. In these three families non-farm income already surpassed farm income.

In contrast, four part-time farming families, because of their larger number of dependents, needed to maintain additional income sources to complement their farm income. In all these families – a young couple with the mother of the wife, a middle-aged couple with four children, and two female-headed families supporting children – the income derived from their agricultural labor was larger than that from cultivation of their own land. Their cash liquidity was the highest of the four categories.

These four categories of group members, having different livelihood strategies, reacted to the interventions of AOFG in different ways (Table 2.2). The fact that all the 13 families joined this farmers' group means that they expected some kind of benefit from participation. Yet expectation varied from one farmer to another, even in the same group. Only the commercial farmer, KJ, was active in both organic farming and Fair Trade coffee. The five subsistence farmers were active

Table 2.2 Different reactions to the two initiatives (Kerala, India)

	Type A 5 full-time subsistence farmers	Type B One commercial farmer	Type C 3 wealthier part-time farmers	Type D 4 agricultural labor-cum-farmers
Active in both organic and Fair Trade		1 member		
Active in organic, inactive in Fair Trade	5 members			2 members
Active in Fair Trade, inactive in organic			3 members	
Inactive in both				2 members

(13 member households in total)

in adopting the new organic method but inactive in selling coffee in the desig-nated way. On the contrary, the three wealthier part-time farmers were active in selling coffee to AOFG but inactive in practicing the new organic method. Finally, responses of the laborer-cum-farmer families were divided: two families were active in organic farming but inactive in Fair Trade coffee; the other two were active in neither. Further analysis focuses on uncovering reasons for such different reactions of the four types of farmers.

2.4 The two initiatives incorporated in different livelihood strategies

2.4.1 Type A: Crop diversification rather than a single Fair Trade crop

The objective of the five full-time subsistence farmers was to maintain their mod-est living. Therefore, they gave priority to ensuring an income throughout the year rather than to maximizing annual income. The small and marginal farmers planted a variety of crops with different harvest periods. Cocoa was one of their preferred crops, which was harvested and sold once a week throughout the year. The most marginal farmer in this group, owning only 0.5 acres in total, expressed his ambivalence:

> It is attractive to sell coffee to AOFG at the higher-than-market price, but in such tiny plots of my land, there is no more space for increasing coffee bushes. Coffee brings income only once a year. I do not want to decrease space for other crops.

Although they sold some of their harvests from currently available coffee bushes to AOFG, they did not intend to buy any seedlings. Furthermore, there were other reasons why they were not embracing the Fair Trade coffee.

Before AOFG's intervention in coffee, the world coffee price fell below small farmers' costs of production in 1999 and hit an all-time low in real terms in 2001 (Jaffee, 2007, pp. 42–49). This situation continued until 2004. During this crisis period, many farmers in this area reduced their coffee bushes and planted rubber as a new alternative cash crop. Four of the five subsistence farm-ers had rubber trees as well. Two of them, free of charge, had received rubber saplings with chemical fertilizer from the Rubber Board. Although they had to wait for the first harvest of rubber latex for a few years more, the recent good producer prices of rubber allowed them to look forward to future profits.[5] Rubber was also attractive because of its year-round income-generating nature. The interest of many local farmers may have shifted from coffee to rubber. As a result, about 25 percent of coffee seedlings in AOFG's nurseries remained unsold.

For such subsistence farmers, it was also important to reduce production costs and to increase profits on the premise of access to year-round income. Because all five members were fairly elderly, they were required to hire laborers for coffee harvesting. The oldest farmer had already decided to lease out all his coffee bushes from 2008. The group president, named VG, in his mid-40s, talked about his new plan:

> I have recently purchased nutmeg seedlings from another farmer. I intend to convert some coffee into nutmeg. Nutmeg is easier than coffee in harvesting. So, I will be able to harvest by myself when I become older. Thus, the conversion will reduce my production cost.

To cover labor costs, they also tried to maximize cash income from each harvest. AOFG offered the best price only for fresh ripe berries for the future Fairtrade product, which allowed subsistence farmers to gain more cash income from local markets by drying and husking berries themselves before sales.

The final reason why the subsistence farmers were indifferent to the Fair Trade coffee is that coffee was an important cashable asset. One of them was not able to sell any coffee berries to AOFG in this season:

> I leased out all my coffee bushes to a local trader this year. To pay for a medical bill, I received the lease fee in cash four months before the harvest season. I have taken care of the bushes as usual, but all the harvest goes to the trader this year.

Leasing out coffee bushes was a strategy typical of poor farmers when they needed a large amount of cash urgently. It was their only way of paying medical bills or dowries for the marriage of their daughters without falling into debt. Therefore, they were not able to promise to constantly supply their coffee berries to AOFG.

Instead, it was the new organic method introduced by AOFG that encouraged the five members to join this group. One of them had never used chemical fertilizers, drawing on cow dung and crop residues, but worried about a disease observed on her cocoa. She had started to use the new bio-fertilizer and had gradually seen a positive effect. Another two did not want to spend much money on fertilizers, but used a chemical fertilizer for rubber saplings delivered by the Rubber Board. After the five-year period of free fertilizer distribution ended, they again became naturally organic farmers. For these traditionally organic farmers, the new organic method was a way of improving their farming at a small cost. Another subsistence farmer, who had used both chemical fertilizers and organic manures, regarded the bio-fertilizer as a cheaper substitute for chemicals. These subsistence farmers were not so keen on investing in agriculture, that is, increasing production with chemicals, nor were they satisfied with traditional organic farming. Despite their limited knowledge of organic certification, therefore, they were interested in improving farming organically.

Unlike the other four members, the group president, VG, was enthusiastic about farming itself and had experimented with many methods of fertilizing, both conventional and organic. As soon as AOFG introduced the bio-fertilizer to the group, he started to use it for many crops experimentally:

> I have already confirmed some positive effects of the bio-fertilizer on all my four plots. But it will take two years to learn the most effective way of using this bio-fertilizer. I still use chemical fertilizer only for rubber, which cannot grow sufficiently with organic manures alone. The introduced bio-fertilizer is not a panacea for all crops. I want AOFG to introduce more drastic organic methods.

The increasing concern about rubber, a promising new cash crop, challenged the conversion of the association members into 100 percent organic farmers.

2.4.2 Type B: Making the most of the unofficially encroached land

KJ, who was officially a small farmer but unofficially a large commercial farmer, attempted to take advantage of many aspects of the farmers' association through his participation. His family's first reason for participation was to gain *group* organic certification. Because the soil of the encroached forestland had been sufficiently fertile, KJ did not have to use any fertilizer. Although coffee cultivated in the forest was completely organic, it was impossible to obtain individual certification for the unregistered land. Pursuing group organic certification for the farmers' association, KJ expected to sell coffee from the unregistered land as certified organic. This family was therefore keen to facilitate other members' adoption of the new organic method. The family offered to the farmers' association a part of their homestead and a supply of electricity that enabled the association to operate a refiner for the bio-fertilizer there (Figure 2.1). Member farmers, not only from KJ's group but also from other groups, came to his homestead to purchase the liquid fertilizer.

KJ also showed a great interest in selling part of his coffee to Fair Trade markets. Being typical of large-scale producers, he was concerned about selling coffee at higher prices as well as diversifying sales channels. Unlike other farmers, who tended to sell coffee berries immediately after harvest or primary processing, he stored dried berries for several months, waiting for the best timing. He was not interested in rubber because the forestland was unsuitable for its growth, since rubber needed more sunlight than coffee does.

The third reason for his participation was that he expected to receive technical and financial services for his coffee cultivation through the association or group. For the unregistered land he was never allowed to receive any subsidy or service directly from government offices. He had already purchased 300 coffee seedlings originally from the Coffee Board through AOFG at a subsidized price.

2.4.3 Type C: Compatibility with more lucrative non-farm activities

The houses of the three part-time farmer members showed evidence of their higher standard of living. While their non-farm income was larger than their farm income, all of them intended to keep farming their own land as a basic asset. In these families both husbands and wives were too busy with their non-farm activities to spare much time for farming. Although the son in one family was a full-time farmer at that time, he was also seeking an opportunity to work as a professional driver. Therefore, these families' concern in farming was to maximize profits for the smallest possible investment of time. One family was leasing out coffee and rubber plants; another did the same with their paddy field.

For these families it was convenient to sell fresh coffee berries to AOFG without any postharvest activities. They appreciated the better price offered by AOFG, and they saved time and transport costs that they otherwise would have incurred. They were willing to cooperate with AOFG in supplying their coffee berries for the Fair Trade–processed coffee. Two of the families had purchased new coffee seedlings from AOFG.

It is usually difficult for busy part-time farmers to meet the higher labor demands for organic production (e.g., Thavat, 2011, p. 292). The three members were no exception. Before knowing AOFG they relied more on chemical fertilizers than they did on organic manures, although – being comparatively more highly educated – they understood the negative consequences of chemicals. When AOFG introduced the new bio-fertilizer, its easy preparation immediately attracted them. However, they gradually found it inconvenient to use this liquid fertilizer. A quantity of the bio-fertilizer, adequate to the capacity of the refiner, was produced on fixed dates. The woman who ran a cafeteria missed the latest opportunity to buy the bio-fertilizer:

> We were informed beforehand of the production schedule. But on the day, my husband worked in the town all the day (as a load carrier) and I also had to go out of the village. Even if I could buy, I could not find time to apply the liquid within 24 hours after the purchase. I wish I could buy and apply it whenever I like.

It was difficult for the busy part-time farmers to comply with AOFG's bio-fertilizer production schedule. Consequently, they used the bio-fertilizer less frequently. They also believed that chemical fertilizers were indispensable for rubber cultivation.

2.4.4 Type D: A transition from agricultural laborers to subsistence farmers

Although laborer-cum-farmers needed to earn wages as laborers, they also took care of their small parcels of land as earnestly as full-time subsistence farmers did. A primary difference observed between laborer-cum-farmers and full-time subsistence farmers was the larger number of dependents in the former. When their children graduate from school, laborer-cum-farmers are highly likely to become full-time subsistence farmers. Like subsistence farmers, they also made efforts to secure year-round income by planting a variety of trees and crops. Therefore, it was physically difficult for them to increase the number of coffee plants. All the four laborer-cum-farmer families in this group were reluctant to sell fresh coffee berries to AOFG immediately after harvest. Although they were busy working for other farms, they also tried to maximize cash income from their own land. They inevitably preferred to sell dried and husked beans to local markets at higher prices.

While one female-headed family had never used chemicals to reduce production costs, three of the laborer-cum-farmer members tended to use more chemicals than did the full-time subsistence farmers. The fact that they were busier and poorer than full-time farmers may have encouraged them to use chemicals, both to save time and to maximize production. When the bio-fertilizer was introduced, two of them, LS and TK, immediately switched to this organic fertilizer. In addition to the cheaper cost of the bio-fertilizer and its easy preparation, LS described another merit of the bio-fertilizer as follows:

> When I apply chemical fertilizer, I have to carefully measure a proper quantity for each crop. In applying this bio-fertilizer, I do not have to be nervous. I can use this fertilizer for all crops. It does not damage crops even if I apply too much.

TK explained why many farmers had quickly adopted this bio-fertilizer in this area:

> The most popular item of chemical fertilizer became unavailable in the local market about one year ago. So, we had to give up the use of chemicals. It was timely that AOFG introduced a good alternative to us.

This narrative was endorsed by fertilizer retailers in the nearest town. The most popular and cheapest item subsidized by the government had not been available for the past year because of a sharp drop in production. According to the retailers, the sales of organic fertilizers had increased since then.

The rest, two of the laborer-cum-farmer members, seemed indifferent to the new organic method introduced by AOFG. The female head of family, traditionally organic, confessed that even the bio-fertilizer was more costly than traditional manures. These two members also tended to be absent from the group's regular meetings. They may have joined this group to maintain good relationships with the neighbors, who were also their employers. More persuasively, a group savings account that they expected to have in a commercial bank may have motivated them to participate in the farmers' association. Such title-less households will comply with the Fairtrade and organic regulations as long as they have access to bank loans through the group.

2.5 Comparison with the theoretical double certification

When AOFG started this project, they expected two theoretical merits of double certification. First, Fair Trade helps producers to perceive the market value of organic agriculture specifically; second, double certification increases accessible markets. We compare the findings from the case study with these theoretical merits of double certification.

AOFG decided to pursue a Fairtrade label to enable member farmers to perceive the market value of organic agriculture, because Fairtrade certification was more oriented to export than was organic certification. This attempt was half successful. In this project the members have learned why organic products can be sold at higher prices through AOFG's efforts to purchase organic coffee from them for the future Fairtrade market. However, the market value of organic agriculture was not necessarily linked with the Fair Trade coffee in this case. Rather, the members expected to sell many other crops – such as vegetables and local nuts – at higher-than-conventional market prices through AOFG. Although the members understood the market value of organic agriculture, AOFG did not anticipate such dependency on the project.

In the coffee sector the global organic market is larger than the Fair Trade market (Raynolds, Murray, and Heller, 2007). Double certification enables disadvantaged producers to access both markets. Organic certification further opens up opportunities to export more items in response to changing market demands. This was the original intention of AOFG. However, ironically, only one member household in the focus group reacted positively to double certification, a substantially large commercial farmer who could afford to diversify the sales channels of a single item – coffee in this case – by adding the organic and Fair Trade values to the item.

The majority of marginal and small farmers accepted organic certification, which covers any crop, but they were not so interested in Fairtrade certification that focused on a single crop. For them, diversification of crops was more important than was diversification of sales channels. The organic initiative meets small and marginal farmers' wishes to gain better profits throughout a year and to reduce costs for chemicals. Even if they do not accept the organic method introduced by AOFG, they can enjoy the added value as long as their products are organic. While some of the members still used chemical fertilizers for their important cash crops, they also experimented to observe the effects of the new bio-fertilizer with minor or decaying crops. There was a possibility that the outcome of such experiments may gradually change them from 50 percent to 100 percent organic farmers.

Conversely, a focus on one single product required by Fair Trade, unfavorable to full-time subsistence farmers, was welcomed by the wealthier part-time farmers. Because they had already diversified their income sources, they were less concerned about further diversification in one of the sources, farming. Rather, they were concerned with making more profits without spending time. The concern to save time may be more suited to Fairtrade certification, which requires only basic environmental criteria (Raynolds et al., 2007, p. 154), rather than to organic certification, which requires a farm to be completely organic.

This case does not allow us to judge whether the confluence of the two initiatives is more helpful to producers than either one of the two initiatives individually, because the confluence itself depends on each producer's livelihood strategy. Some strategies make it difficult to adopt both opportunities in parallel. The existence of different stakeholders within a single group predicts difficulty in organizing them for common objectives. Those who intervene in the organization process, like AOFG, should be sensitive to the fact that not all small farmers necessarily appreciate the confluence of the two initiatives. A failure to do so will expose the two initiatives to the danger of being used for unexpected purposes far from the original ones. It is necessary, albeit difficult in reality, to detect and eliminate officially small but substantially large farmers. In response to a recently increasing shift from farm to non-farm activities in the rural South (e.g., Rigg, 2006), a separate association for part-time farmers may be required, operating independently of an association for full-time farmers.

In this case the fundamental difference between the two certification systems appeared as a contradiction between how to grow a variety of crops through organic means and how to gain from a specific crop through Fair Trade. This contradiction can be interpreted not only as that between the organic and Fair Trade certifications but also as that between the organic movement and Fair Trade certification. While the difference between the two certification systems in what is certified is the very contradiction between the two systems, subsistence farmers responded more directly to the essence of the organic movement, that is, "agro-ecosystem health," especially "biodiversity" (FAO, 2014). Although the NGO project introduced the organic movement to small farmers, the project's focus on coffee obviously required the farmers to adjust their production to the market-oriented certification, without conveying the essence of the Fair Trade movement.

2.6 Concluding remarks

This chapter has focused on the difference of the two initiatives at the production level: Fair Trade focuses on specific products, and organic agriculture targets production units. How do producers in the South cope with this discrepancy? The case study in South India revealed that even a single group of smallholders contained different stakeholders who had different livelihood strategies and thus reacted differently to the two initiatives. Some members were more oriented toward Fair Trade; some toward organic agriculture. Some were more attracted by other interventions provided through the farmers' association. The existence of different stakeholders within a single group predicted difficulty in organizing them for common objectives. The two initiatives did not necessarily reinforce each other.

The contradiction identified in this case eventually compelled the NGO project to change its nature. In the author's second visit to this project of AOFG two years later, she noticed that the number of organized groups increased from 52 to 84 and that, in addition, AFOG was attempting to organize two more farmers' associations in other parts of highland Kerala for the same export purposes. However, the expansion of the project did not mean the success of the project.

While 6 out of 13 member households in the focus group sold coffee to AOFG in the harvest season of 2008–2009, none of them sold coffee in 2009–2010.[6] This phenomenon was not unique to this group, but was common to most of the early organized 52 groups. In the 2009–2010 harvest season, surprisingly, 99 percent of the total coffee sold to AOFG came from new groups. The new members were more dependent on coffee production. Because the geographic conditions of the areas were not suitable for rubber plantation, the new members were not interested in replacing coffee with rubber. The majority of the new members had land ownership and access to bank loans. Their average land size was obviously larger than that of original members, and their house conditions looked much better.

The expansion of the project obviously showed that the objective of the project had changed: from pursuing both certifications to obtaining Fairtrade certification first. Although the majority of the subsistence farmers highly evaluated organic farming, within a limited budget for the project it was actually difficult to convert all members into 100 percent organic farmers. It is, in theory, easier to acquire Fairtrade certification, which requires only basic environmental criteria (Raynolds et al., 2007, p. 154). Participation in the Fair Trade networks necessarily demands the stable supply of the target product of export quality. AOFG seems to have in haste organized the new groups for coffee supply. AOFG did not specially promote organic farming for the new groups any longer. Arguably, the project shifted its focus from promoting organic farming for poor farmers – the confluence of organic and Fair Trade movements – to meeting requirements for Fairtrade certification. The attempt of double certification in this case paradoxically ended in the exclusion of small and marginal farmers who *should* benefit from the two movements.

Notes

1 This labeling system is unique to the agricultural sector. In the case of handcrafts, a Fair Trade label is given not to specific items but to producer groups who can sell a variety of products.
2 According to AOFG's experiment, the most popular and cheapest chemical fertilizer cost Rs. 984 to 1312 per acre each year; the bio-fertilizer cost Rs. 600 per acre in the first year and Rs. 400 per acre in the second year and each year thereafter. The retail prices in December 2008 were applied for calculation. Rs. 48 was equivalent to US$1.00.
3 In the 2008 harvest season AOFG paid Rs. 20/kg, against Rs. 15/kg on the local market.
4 Concerning some coffee varieties, the speed of ripening differs among berries on a bush, which makes it difficult for farmers to pick out only ripe berries after harvest – but if they wait until all berries are ripe, some are by then rotten. Therefore, AOFG introduced a specific variety whose berries ripened evenly.
5 After planting a rubber sapling, a farmer usually takes about eight years to gain the first income. According to local wholesalers, the Olympic Games in China boosted the price of rubber in 2008.
6 Coffee was harvested from late November to January in the study area.

3 Fair Trade, organic, and genetically modified organisms

A case from Central India

The implications of the convergence of Fair Trade and organic initiatives can vary from one crop to another and from one context to another. In this chapter we pay attention to a notable context in which the two initiatives meet: the cotton sector. Even if the two initiatives look similar, they may show a difference when confronted with a third factor. The cotton sector provides such a third factor that allows us to observe how differently the respective initiatives react to the third factor.

3.1 A contradiction between the Fair Trade and organic movements

In cotton cultivation, the use of genetically modified (GM) or *Bacillus thuringiensis* (Bt) hybrid seeds has been controversial since the commercialization of the first Bt cotton varieties in 1990 (Baffes, 2011, p. 3). Bt seeds have been introduced to increase cotton yields as well as to reduce the use of chemical inputs (see Tripp, 2009). This innovation is generally welcomed by cotton farmers in the global South, which reminds us that the Green Revolution once spread rapidly among Southern farmers (e.g., Stone, 2007). However, Bassett (2010, p. 52) anticipates that "Bt contamination of non-GM cotton through cross-pollination jeopardizes both the organic and fair-trade programs." Organic certifications clearly ban the use of GM seeds (IFOAM, 2007). The Fairtrade standard set by Fairtrade International (FLO), which was originally intended to help disadvantaged farmers improve their livelihoods, also prohibits Fairtrade-certified producer organizations from using GM seeds, albeit not as definitively as organic certifications do (Fairtrade Foundation, 2011). According to the Fairtrade Foundation (2011), Fairtrade products are *not* guaranteed to be GM free:

> Fairtrade does not test harvested crops for GM traits for two reasons. First, the costs would be prohibitive in relation to the potential risk. Secondly, if a Fairtrade-certified farm is accidentally contaminated by pollen from GM crops that crop would then have to be excluded from the Fairtrade market. We believe it would be unfair to punish farmers for events like this that are beyond their control.

It is possible to interpret this statement as an excuse for the intrusion of Bt seeds into Fairtrade cotton. Although the organic initiative can judge the propriety of GM organisms only from the ecological viewpoint, the Fair Trade initiative may not be able to neglect positive impacts of Bt seeds on poor farmers and laborers verified by scholars such as Ramasundaram, Vennila, and Ingle (2007), Subramanian and Qaim (2009, 2010), and Qaim (2010) as long as Fair Trade gives priority to its contribution to poverty reduction. This slight but significant difference between the two initiatives in terms of their attitudes toward GM crops seems to make them incompatible in real production settings. When a new technology such as Bt seeds is introduced, how do small farmers choose between the innovation, Fair Trade, and organic farming? How should we interpret the confluence of Fair Trade and organic initiatives in the context of the innovation-influenced cotton sector?

To explore these questions, this chapter applies Guthman's (2004a, 2004b) conventionalization thesis on organic farming in the global North to the Southern context and draws on a case study conducted in a major cotton-growing country, India. As detailed in the next section, Guthman argues that agribusiness involvement has made organic farming more shallow or industrial, "effectively lessening some of the distinctiveness of organic vis-à-vis conventional farming" (Guthman, 2004b, p. 307). The next section frames the application of this conventionalization thesis on the basis of a literature review to interpret the confluence of the Fair Trade and organic initiatives, as well as plausible interactions between the two. The third section outlines the case for observation, and the fourth section reveals how Fairtrade-certified small farmers adopt or do not adopt Bt seeds and organic farming in a real setting. By applying the theoretical framework to the interpretation of data from the case study, the fifth section identifies a paradox of the two initiatives as an outcome of the confluence. The final section briefly summarizes major findings and policy implications from the case study.

3.2 Theoretical propositions

The Fair Trade and organic initiatives are, in reality, practiced through the certification systems. Although the certification systems do not necessarily convey the original purposes of the two initiatives (e.g., Getz and Shreck, 2006; Jaffee and Howard, 2010), observation of the confluence of the two initiatives needs to draw primarily on the *certified* initiatives. Whereas the respective certification systems for the two initiatives were designed commonly for Northern consumers, there is a clear distinction in terms of motive to purchase: Fair Trade was invented and is supported for the benefit of disadvantaged producers in the global South, and the organic certification is primarily for the benefit of Northern consumers (Bowes and Croft, 2007, pp. 278–279; Gonzalez and Nigh, 2005, p. 450). Therefore, it is in the first place necessary to clarify the meaning of organic farming for small producers in the South. The rest of this section attempts to theorize plausible interactions between the two initiatives on the part of Southern producers, also referring to the influence of Bt cotton as a new technology.

3.2.1 *Organic farming introduced to the global South*

Although farmers in the global South have traditionally operated organic farming, or, more precisely, natural farming, the value of organic practices was realized and defined in the global North (Raynolds, 2004). The so-called "conventionalization debate," starting from Buck, Getz, and Guthman (1997), is central to the organic discourse in the North. Guthman (2004a) documents how agribusiness corporations have permeated the organic sector and how the conventional industrial model has been replicated in organics, enabling such corporations to capture more power and profit. This process is called "conventionalization." In other words, the principle of organic farming – that ecologically healthy farm systems promote agriculture – has been gradually transformed into an alternative means of capital accumulation through organic regulation and industrial agriculture. As a result, "agrarian dreams" such as working with nature and small-scale production within localized economies have rarely been realized by producers (Guthman, 2002, 2004a). Studies by Smith and Marsden (2004), Lockie and Halpin (2005), De Wit and Verhoog (2007), Best (2008), and Guptill (2009) basically support this conventionalization thesis, albeit finding the existence of non-conventionalized, pro-environment farmers as well. Others primarily criticize the simple linear trajectory of the conventionalization thesis with divergent empirical evidence in different settings (Coombes and Campbell, 1998; Hall and Mogyrody, 2001; Pratt, 2009).

Although the conventionalization debate has not yet been concluded, when the organic movement entered the global South, people had "agrarian dreams" again. The organic movement can theoretically benefit Southern producers in two ways. First, their small-scale natural farming with unsophisticated technologies may be evaluated as a positive sales point by Northern buyers and consumers (Nigh, 1997). Second, organic farming may revitalize both the soil's and farmers' health, which have been exhausted and harmed by conventional farming with chemical inputs under the Green Revolution (Shiva, 1991). Many NGOs and some official development agencies promote organic farming, expecting beneficial impacts on both natural resource management and farm incomes (McNeely and Scherr, 2003; Scialabba and Hattam, 2002).

Because organic markets have not yet been developed in the South, Southern organic producers have from the beginning sought access to Northern markets (Wynen, 2003, pp. 214–216). In terms of entry to Northern markets, certification plays a significant role. Inevitably, "producers in Latin America, Africa, and Asia have joined with exporters and certifying organizations to form organic trade associations which work with Northern distributors to create South-North trade circuits"; organic trade associations in the South have been required to support local certification systems that adhere to Northern importing standards (Raynolds, 2004, p. 730). As a result, Nelson, Gomez Tovar, Rindermann, and Gomez Cruz (2010) view mainstream organic certification as a means of increasing the dependency of Southern producers on rich consumers and buyers in the North. In other words, the organic concept in the South was easily linked with the organic concept in the North as "the alternative accumulation strategy for agrarian capitalism" rather than as "an alternative framework for working the rural environment" (Woods, 2011, p. 89).[1]

Does the involvement of Southern small farmers in the Northern conventionalization process bring any benefits to them? Organic agriculture is originally defined as "a production system that sustains the health of soils, ecosystems and people, [relying] on ecological processes, biodiversity and cycles adapted to local conditions, rather than the use of inputs with adverse effects" (IFOAM, 2011). As many authors argue (Guthman, 2004b; Mutersbaugh, 2005; Raynolds, 2000), however, the basic tenets of organic agriculture have been gradually solidified in the process of establishing and enforcing standards, resulting in "the simplification of the operational meaning toward a single variable: allowable versus prohibited inputs in organics" (Jaffee and Howard, 2010, p. 397). If organic standards are weakened and simplified, natural farming traditionally operated with emphasis on biodiversity by small and marginal farmers in the South may not be evaluated properly (Eernstman and Wals, 2009). In reality, in the conversion from conventional into certified organic farming, small producers are less advantaged than large producers are (Gomez Tovar, Martin, Gomez Cruz, and Mutersbaugh, 2005; Mutersbaugh, 2002; Raynolds, 2008). Although organic farming itself has a high potential for poverty reduction and more sustainable livelihoods in developing countries (Pretty, Morison, and Hine, 2003; Scialabba and Hattam, 2002), it is not easy for many small farmers, who are the target of the Fair Trade initiative as well, to enjoy the benefits of organic farming under the certification system.

3.2.2 Plausible interactions between Fair Trade and organic initiatives

Albeit Fair Trade and organic initiatives practiced through the respective certification systems have general principles in common (see Chapter 1), as Table 3.1 summarizes, there are clear differences in certification: the organic puts more emphasis on environmental regulations, and Fairtrade on benefits for small producers. When the two certification systems are combined, more benefits and more burdens are anticipated for small farmers. The certification processes interact with each other in two directions: (a) from organic to Fairtrade and (b) from Fairtrade to organic. Interactions may not work equally in both directions because of the differences between the two certifications.

In the former direction (a), organic certification enables organic-certified farmer cooperatives to obtain Fairtrade certification easily because the organic certification guarantees the environmental requirements that Fairtrade producers should meet. As pointed out, however, small producers have less access to organic certification than do large producers. There might not be many cases in which small farmer associations obtain organic certification – demanding a three-year conversion period from all association members – *before* obtaining Fairtrade certification.

Table 3.1 Major differences between Fairtrade and organic certifications

	Fairtrade	Organic	Double
Time required before certification	Acceptance process takes about six months	Two years for annual crops and three years for perennial crops after conversion	Same as organic or longer
Type of producer	Must be smallholders as members of a democratically organized association[a]	Unspecified	Same as Fairtrade
Agro-ecological standards	The use of chemical fertilizers, synthetic pesticides, and genetically engineered planting materials is *restricted*; the use of herbicides and some specified pesticides is *prohibited*	The use of chemical and genetically engineered planting materials, non-organic fertilizers and synthetic herbicides and pesticides is *prohibited*	Same as organic
Social standards	More specific labor conditions are stipulated	Basic labor conditions are stipulated	Same as Fairtrade
Producer prices	Guaranteed minimum above the world price, moving up with the market	Tend to be higher than non-organic, moving up and down with the market	20% higher than the Fairtrade conventional minimum prices[b]
Credit	Up to 60% of the purchase price is given to the producer organizations	Unspecified	Same as Fairtrade
Social premiums	The buyers must pay US$5 cents per kilo for seed cotton	None	Same as Fairtrade
Trade relations	Must be as direct as possible and aimed at long-term trading relations	Unspecified	Same as Fairtrade

Sources: IFOAM (2007); INDOCERT (2011); FLO (2011a, 2011b); Raynolds (2000)

Notes:
[a] There is another set of standards for hired labor in plantations different from that for small producer organizations.
[b] This is the case for cotton. The Fairtrade organic premium varies according to the product.

In the latter direction (b), although Fairtrade certification by itself does not enable Fairtrade-certified farmers to obtain an organic certification, Fairtrade may contribute to organic farming in two ways. First, Fairtrade certification is sometimes obtained strategically as a first step to organic certification. In the process of pursuing organic certification, Fairtrade certification helps producers perceive the monetary value of organics specifically and tangibly by enabling them to sell a specific Fairtrade product (see Chapter 2). Owing to the absence of local organic markets in most developing countries, farmers usually cannot realize the organic value before obtaining an organic certification. By accessing the Fairtrade market – the most easily accessible export market for small farmers in the South – they may be able to perceive the organic value. In brief, Fairtrade certification can be used as an incentive to organic farming. Second, as Jaffee (2007, pp. 138–164), Bassett (2010), and Frundt (2009, pp. 80–81) respectively report in the cases of coffee, cotton, and bananas, the Fair Trade certification system tends to promote the diffusion of organic farming technologies through the demonstration effect of certified association members' practices. Fairtrade certification may encourage Fairtrade and non-Fairtrade conventional farmers to convert into organic farmers even if they are not officially certified as organic.

What outcomes do such interactions result in? As mentioned before, it is possible to view the organic initiative in the global South as an exploitation of small producers because of Northern (and Southern) agribusiness players' alternative accumulation strategy for agrarian capitalism; the tendency to weaken standards for organics seems to intensify this strategy, which eventually hinders small farmers from taking advantage of organics. When combined with the Fair Trade initiative, does the nature of such organics change? Jaffee and Howard (2010, p. 397) also argue, however, that through the process of establishing and enforcing standards the operational meaning of Fair Trade, like that of organics, has been reduced to a single variable: "Payment of a minimum price in fair trade, the level of which is no longer linked to actual family livelihoods." If the Fair Trade initiative does not give sufficient power to certified farmers, does the organic initiative contribute to agrarian capitalism, regardless of the interactions between the two?

Whether the organic concept as an alternative means of capital accumulation is promoted or not eventually depends on the context regarding the confluence of the two initiatives. In theory, the confluence has both possibilities of strengthening the organic concept as "an alternative accumulation strategy for agrarian capitalism" and of eliciting the original nature of organics as "an alternative framework for working the rural environment" (Woods, 2011, p. 89). Drawing on these two contrasting concepts as a framework for interpreting the confluence of the Fair Trade and organic initiatives, this chapter observes how the organic concept varies in response to a context.

3.2.3 Bt cotton as a given context

The context given here is the cotton sector, in which the introduction of Bt seeds has aroused vigorous controversy for the last decade (e.g., Baffes, 2011). On one hand, anti-GM organism movements are reported from India, South Africa, and Brazil (Herring, 2005; Scoones, 2008), and there are authors who argue that Bt cotton may not be economically viable without accessible support services – input supplies, technical advice, and finance – and reliable output markets (Gouse, Shankar, and Thirtle, 2008), or in the long term because of the increased need for pesticides to cope with the emergence of secondary pests or the high levels of pest resistance to the Bt toxin (Dowd-Uribe and Bingen, 2008). On the other hand, scientific and policy consensus regards GM crop technology as a "pro-poor" technology that contributes to agricultural and economic development (Glover, 2010; Scoones, 2002). In fact, the use of GM cotton seeds has spread rapidly in the global North and South. The world's total area of Bt cotton increased from 0.8 million hectares in 1996 to 13.4 million hectares in 2006 (GMO Compass, 2011) and Bt cotton accounted for an estimated 52 percent of the world's total cotton-cultivated area in 2009–2010 (Baffes, 2011, p. 14). As Srinivas (2002) argues, technological and economic factors may outweigh environmental concerns.

Whether it is pro-poor or not, this new biotechnology has doubtless worked as an accumulation strategy for seed companies. "Even when transgenic varieties are developed by public research, . . . the seed will be delivered by private seed companies" (Tripp, 2001, p. 259). Expressing consequent ethical concern about GM crops, many authors point to Northern or transnational corporations' excessive dominance and control of Southern markets (Gibbs, 2000; Newell, 2009; Reece, 2006; Srinivasan, 2003; Weale, 2010; Zerbe, 2004). In the context of Indian cotton, Shah (2005) regards the spread of Bt seeds as the result of the alliance between local and global elites. Such elites include not only transnational corporations but also local seed companies supplying "pirate" or "illegal" seeds and wealthy farmers with access to land and water. According to these perspectives, small farmers are exploited by local and global elites' accumulation strategies.

Bt cotton can affect small farmers in two aspects: as a new technology and as a new cash crop. First, those who had been suffering from pest attacks with non-Bt seeds adopted Bt seeds as a newly introduced technology for alleviating the pest problem (Roy, Herring, and Geisler, 2007). Second, farmers who had no experience in cotton cultivation hitherto were attracted by higher yields from Bt seeds and began cotton cultivation with them (Stone, 2007, p. 76). When organic farming is introduced to a cotton-growing area, the former group of farmers are required to make a decision as to whether they should switch from conventional farming with Bt seeds to organic farming with non-Bt seeds.[2] The second group of farmers could have two options, conventional and organic, at the same time. Each farmer's experience before the introduction of organic farming will influence his or her decision on the adoption of organic cotton cultivation.

Conversely, the Fair Trade movement appears to be substantially free from the controversies surrounding Bt seeds for two reasons. One reason is that Fairtrade certification does not strictly examine Bt contamination, as mentioned in the beginning of this chapter. The other is that Fairtrade certification itself does not offer any equivalent to an organic price premium that testifies to the GM-free nature of the product but guarantees the minimum farm-gate price, regardless of local price trends that may reflect the prevalence of Bt seeds. Given the spread of Bt cotton, a case of small farmers in central India is studied to see how they perform with regard to the confluence of Fair Trade and organic initiatives.

3.3 Outline of the case study

3.3.1 Cotton cultivation in India

India is the second-largest cotton producer and consumer in the world (Osakwe, 2009). Three states, Gujarat, Maharashtra, and Andhra Pradesh, account for about 70 percent of the total cotton production (Osakwe, 2009).[3] In the world cotton sector, India, albeit having the largest area under cotton cultivation, is known for its low productivity because of severe pest ravages and its predominant cultivation under rain-fed conditions (Eyhorn, 2007, p. 24; Narayanamoorthy and Kalamkar, 2006, p. 2716). The inevitable use of pesticides not only increases the financial burden of the farmers but also creates health hazards and environmental risks; such financial burden is related to a high incidence of poor farmers' suicides in the cotton-growing areas (Narayanamoorthy and Kalamkar, 2006, p. 2716; Patil, 2002). The introduction of Bt seeds was expected to improve this situation. The national average yield of cotton dramatically increased from 308 kg per hectare in 2001–2002 to 568 kg per hectare in the 2009–2010 season[4]: the remarkable increase in yield is mainly attributed to Bt seeds that were first commercialized in 2002 (Choudhary and Gaur, 2010, p. 10).

Bt seeds have spread rapidly: only one company dealt with three Bt hybrids in 2002, but 35 companies received approval to sell 522 Bt hybrids by 2009 (Choudhary and Gaur, 2010, p. 16). Consequently, Bt cotton was estimated to cover 86 percent of the national cotton area in 2010 (Choudhary and Gaur, 2011). As regards the benefits of Bt cotton, in addition to the increase in crop yield and reduction in pesticide use, Ramasundaram et al. (2007) empirically verify an increase in net profit, and analyses by Subramanian and Qaim (2009, 2010) and Qaim (2010) show that Bt cotton generates more employment and increases returns to labor, especially in the case of hired female workers. Bt seeds do not, however, necessarily work as a panacea in all situations. Morse, Bennett, and Ismael (2007) point out that there are distinctions between Bt adopters and non-adopters: adopters tend to specialize more in cotton and spend more money on irrigation. Subramanian and Qaim (2009) themselves concede that the total household income effects of Bt cotton adoption are bigger for larger farmers. Although a study by Narayanamoorthy and Kalamkar (2006) suggests that Bt cotton cultivation is substantially better than cultivation with non-Bt varieties in terms of both productivity and profit, it also reveals that quite a few farmers cultivate Bt cotton without distinguishing between Bt and non-Bt varieties and thus continue to use the same quantity of pesticides as they did in the past. On the basis of their case studies from Gujarat and Maharashtra, Lalitha, Bharat, and Viswanathan (2009, p. 167) also point out that "[d]espite the widespread access to . . . transgenic cotton, there are few mechanisms that allow farmers to learn how to use the new technology as part of a more rational approach to insect control." The lack of pest management suitable for Bt seeds may reduce small farmers' net income. The current dominance of Bt seeds may not mean that all cotton farmers evaluate the effect of Bt seeds properly.

Another countermeasure against the indiscriminate use of pesticides is organic cultivation. Whereas organic cotton accounted for only 1.1 percent of global cotton production in 2009–2010, the global organic cotton market jumped from under US$300 million in 2002 to over $4.3 billion in 2009, and the majority, 81 percent, was produced in India (Truscott, Lizarraga, Nagarajan, Tovignan, and Currin, 2010). In other words, it is estimated that about 3 percent of cotton produced in India was certified as organic in 2009–2010 (Choudhary and Gaur, 2011; Truscott et al., 2010). A case study conducted by Eyhorn (2007) on organic cotton farmers in the state of Madhya Pradesh shows that smallholder organic farming systems can produce similar yields to those of conventional farming after the completion of a transitional period of three to four years. His conclusion is that organic farming is a suitable option, particularly for small and marginal farmers who cannot benefit from Green Revolution technologies; such farmers are, at the same time, most vulnerable to decreased yields and incomes in the initial years and consequently find it most difficult to convert to organic farming.

In today's India, theoretically, four types of cotton cultivation can be observed in parallel: (a) conventional with Bt hybrid seeds, (b) conventional with non-Bt hybrid seeds, (c) certified organic, and (d) non-certified organic.[5] The Fair Trade initiative entered such a cotton sector. Although there are no statistical data showing the scale of Fair Trade cotton production in India, it must be smaller than organic cotton: there were 275,300 organic cotton farmers versus 85,000 Fair Trade cotton farmers in the world in 2009–2010 (International Cotton Advisory Committee [ICAC], 2011). The fact that 30 percent of the world Fair Trade seed cotton production was also organic in 2008–2009 suggests the existence of Fair Trade-cum-organic cotton farmers in India (ICAC, 2011).

3.3.2 Organization of small farmers in Telangana, India

The case study draws on an association of cotton farmers organized in a district close to the Maharashtra border, in the state of Telangana (former Andhra Pradesh).[6] A local Agriculture and Organic Farming Group India (AOFG) has supported the formation of the farmers' association. AOFG started their cotton project with the following vision: "To increase the power, negotiating position and knowledge base of small and marginal farmers in India by eradicating exploitation in agricultural production and supply chains, and by mainstreaming the farmers [into the] economic prosperity of the country" (taken from AOFG's project document). The NGO is interested in organic farming as a means of poverty reduction. It has provided the association's members with technical assistance in organic cotton production and has tried to link the association with ethical markets in the North.

AOFG chose this district because of the high incidence of suicide among cotton farmers living there. In December 2010 about 2,700 small farmers had been organized into eight clusters primarily for the purpose of obtaining Fairtrade and organic certifications. AOFG's record showed 4.6 acres as the members' average landholding for 2010–2011. The livelihoods of farmers depended solely on rainfed farming in the district. The attempt to convert to organic farming began in 2006, and 1,017 members had already survived the three-year conversion period in December 2010. The association was certified as Fairtrade in 2007.

3.3.3 Data collection in a focus village

Primary data were collected by the first author in December 2010 and in February–March 2011, focusing on members living in one village (hereinafter, K village) in one of the eight clusters. This cluster was selected because it had the highest dependence on cotton. The cluster members owned on average 3.4 acres of arable land, smaller than the association average, and about 70 percent of their land was allocated to cotton cultivation. Groups in three villages made up the cluster, but only those in K village had actually used Fairtrade premiums according to the members' plans.[7] In K village there were 3 groups consisting of 35 members in total (13 members, 13 members, and 9 members, respectively). According to a survey the District Rural Development Agency conducted in the year 2003–2004, there were 341 households in this village, and all of them were regarded as below the national poverty line. Another survey conducted by the *mandal* revenue office in 2001 showed that about half of the villagers belonged to scheduled castes or tribes.[8] Although there were no statistical data about the villagers' economic activities, the villagers themselves agreed that there were more than 200 agricultural households cultivating cotton, about 100 households cultivating other crops only, and some landless households equivalent to about 10 percent of the village population. Therefore, the 35 members who had at least attempted organic cotton cultivation were a minority in this village. For an understanding of the reality of cotton farmers involved in the confluence of the two initiatives, primary qualitative data were collected from the 35 members and 13 non-member farmers in the same village for comparison, through in-depth interviews and observation of their farm and off-farm activities.[9] Supplementary interviews were held with some member farmers in other clusters, the staff of AOFG, local cotton buyers, local input traders, and government officials in charge of K village. Although organic farming has a wide range of aspects, this case study pays particular attention to the use of inputs regarded as "minimum organic standards in major markets" (Raynolds, 2000, p. 299).

Because the 35 members belonged to a single community in K village, the author selectively approached non-member neighbors in the same community for comparative purposes. Some non-members she accessed refused to be interviewed, afraid that her collected data would be reported to the government. All 13 non-members interviewed cultivated Bt cotton with chemical fertilizers and pesticides. Although no significant difference was observed between the members and non-members in terms of economic and educational levels, the non-members were more conservative and risk-averse than were the members in terms of changing a significant part of their livelihood – from conventional to organic cotton cultivation.

To our surprise, cotton produced by the association was exported to the Fairtrade market only for the initial three seasons. During the two seasons, 2009–2010 and 2010–2011, member farmers were not able to sell any part of their produce to the marketing division of AOFG but instead sold all harvests, both organic and non-organic, to local traders. Without an organic certification, Fair Trade cotton traders did not offer the association better-than-local market prices for the recent two seasons. Although member farmers were satisfied with the recent trend of high local market prices, those who had just completed the three-year conversion period were disappointed at having missed the organic price premium. Furthermore, it turned out that the three groups in the focus village had not been organized at the earliest stage: under the association, 28 out of the 35 members had experienced the last three seasons of cotton cultivation; 7 members experienced two seasons only. This means that 28 members sold their cotton to the Fairtrade market only once. The remaining 7 members had never experienced Fair Trade sales since joining the association.

3.4 Real approach of small farmers to the Fair Trade and organic initiatives under the prevalence of Bt cotton

3.4.1 Motives of the two initiatives

As Table 3.2 shows, the three groups' members joined the association for a variety of reasons. Although the majority of the members understood what the Fairtrade premium was and what the premium was used for in these groups, only one member referred to the Fair Trade initiative as a motive for joining. Because the local market prices of cotton remained fairly high after they joined the association, they never had an opportunity of enjoying the benefit of a minimum guaranteed price. Unlike the Fairtrade premium that producers had to wait several months for after shipping, the organic price premium delivered at the farm gate strongly motivated farmers to join the association and to practice organic cotton cultivation. The members valued organic farming highly, not only because of its price incentive but also for a variety of other reasons (see Table 3.2). Although 2 of the 35 members had already practiced organic cotton cultivation individually, 33 members started only after joining the association. In the case of K village, the organic initiative attracted small farmers' attention more positively than Fair Trade did.

Table 3.2 Motives for participation in the cotton association (three groups in K village, Telangana, India)

Motives	Number of respondents
1. Access to group saving and loans	12
2. Collective action	6
3. Organic farming for the soil	6
4. Organic farming to save on production costs	6
5. Organic farming for a higher selling price	17
6. Organic farming for the health of self, own family, and animals	3
7. Fairtrade premium	1
8. For starting cotton cultivation	1
9. Expecting some material benefits	4

Source: Field survey

Note: Some members cited more than one motive.

Although the two initiatives were the main attributes of the association, the members expected other, different benefits as well from their participation. Group saving and future access to commercial bank loans through the group saving were particularly attractive to many of the members. This seems to be a reason why they kept their membership even if they could not export to the Fairtrade organic market. Others expected to take collective action against the local government, through the groups or the association, demanding irrigation facilities for the area. We should remember that the high incidence of suicide committed by cotton farmers is attributed not only to heavy expenditure on chemicals but also to the decrease of yields caused by uncertain rain-fed farming, both resulting in heavy indebtedness (Patil, 2002). The success or otherwise of Fair Trade and organic initiatives may depend strongly on other constraints that the initiatives cannot eliminate on their own.

Another important point is that one member clearly answered that he had joined the association to start cotton cultivation. In reality, 5 out of the 35 members did not have experience of cotton cultivation before participating in the association, although they did not mention cotton cultivation itself as a motive for participation. Some farmers may have been induced to try cotton cultivation not by Bt seeds but by the Fair Trade and organic initiatives.

3.4.2 Bt seeds as a matter of course

A notable finding from this case study is that many farmers, both members and non-members, in K village were not cognizant of the difference between Bt and non-Bt seeds. The majority of farmers started cotton farming only after official Bt seeds were released; local input-trading shops that mushroomed for the last three or four years in the nearest town had dealt with Bt seeds only. It was physically difficult to buy non-Bt seeds. For the majority of farmers, except the association members, Bt seeds were synonymous with cotton seeds. A few farmers who had cultivated cotton for more than 10 years switched from non-Bt to Bt seeds simply because the local input shops changed their merchandise. Although most farmers realized better yields from Bt seeds, they rarely had knowledge of GM organisms.[10] In brief, Bt cotton was not a special innovation but the only available option for cotton producers outside the AOFG project.

For organic cotton cultivation AOFG distributed non-Bt hybrid seeds only to members of the association. Membership in the association afforded the only opportunity to obtain non-Bt hybrid seeds in the locality.[11] AOFG provided members with non-Bt seeds at the price of 435 rupees (Rs.) per packet, which was lower than the average retail price thanks to bulk purchase by the association. This price was also appealing to some non-member farmers who suffered as a result of the high price of Bt seeds, which cost an average of Rs. 750 per packet. A few of the members confessed to the author that they sometimes resold non-Bt seeds they had bought through AOFG to non-member neighbors to gain a small profit margin.[12] Although AOFG used the number of non-Bt seeds packets distributed to each member as an indication of the acreage under organic cotton cultivation, the existence of such resold seeds hid the real organic practice (Figure 3.1).[13]

Even if most of the members were interested in organic cotton cultivation, they did not adopt organic practices all at once. As Table 3.3 shows, in the first season after joining the association only 14 out of the 35 members cultivated all their cotton organically. Fifteen members cultivated part of their land organically,

Figure 3.1 A non-Bt seeds packet distributed by AOFG (left) and a Bt seeds packet (right) (Telangana, India)

keeping the other parts under conventional tillage. This action came from their risk-averse strategy: they were afraid of reducing yields suddenly by full-scale conversion to organic. The remaining six members tried to cultivate at least part of their land organically at the beginning of the season but were not able to restrain themselves from using chemical pesticides when insects increased in the rainy season.

As observed elsewhere (Eyhorn, 2007), it was not easy for the members to survive low yields during the conversion period. Only eight members were able to continue organic cotton cultivation after participation (see Table 3.4). Some members again started to use Bt seeds with chemical inputs on part of their land, and some completely gave up organic cotton cultivation. Many of those who started to cultivate cotton in both conventional and organic ways decreased the acreage for organic and increased that for conventional cultivation. Although two of the members who did not practice organic methods in the first season adopted organic methods in the second or third season after witnessing the other

Table 3.3 Patterns of cotton cultivation by the focus group members in the *first* season (three groups in K village, Telangana, India)

Patterns	*Number of members*
1. Organic with non-Bt seeds only	14
2. Organic with non-Bt and conventional with Bt in parallel	15
3. Conventional with Bt and non-Bt seeds*	6
Total	35

Source: Field survey

Notes: *Includes one member who leased out all his land plots.

Table 3.4 Patterns of change in cotton cultivation by the focus group members (three groups in K village, Telangana, India)

Patterns[a]	*Number of members*
1-A. Continued organic only	8
1-B. From organic to both organic and conventional	3
1-C. From organic to conventional	3
2. Continued both organic and conventional	15
3-A. Continued conventional only[b]	4
3-B. From conventional to both organic and conventional	2
3-C. From conventional to organic	0
Total	35

Source: Field survey

Notes:
[a] The three patterns in Table 3.3 changed into six different patterns.
[b] Includes one member who leased out all his land plots.

members' sales to the Fairtrade market, the majority of members gradually lost interest in organic cotton cultivation. In correspondence with these changes, AOFG's official record of the sales of non-Bt seeds to the three groups showed a decrease from 69 packets in the 2008–2009 season to 35 packets in the 2010–2011 season. Under these circumstances, although Fairtrade certification could be maintained, it was unlikely that the association would be certified as organic in the near future.

3.4.3 Two dimensions of organic farming: Tradition and innovation

Organic farming itself was nothing new, either to the group members or to non-member farmers in this locality. No respondents, neither members nor non-members, had ever used chemical fertilizers on their farms before starting cotton cultivation. After cotton came to the village as a new cash crop, they began to apply chemical fertilizers only for cotton and red gram (pigeon pea) planted with cotton as a refuge crop in the same plot,[14] continuing to cultivate other food crops such as sorghum and vegetables organically. Therefore, they had fully understood the ecological value and practical techniques of organic farming; they wanted to make their farming as organic as possible.

Concerning cash crops such as cotton, however, economic value is liable to surpass ecological value. The importance attached to economic value was more obvious in relation to leased-in land. Even members who adopted organic practices on their own land usually applied chemical inputs with Bt seeds on leased-in land. Such farmers gave priority to the maximization of profits from the leased-in land with more reliable technologies. A member living in K village explained his strategy:

> The lease period is one season. During the limited time I paid for, I want to harvest as much cotton as possible with Bt seeds. But on my own land, it is important to maintain the soil fertility by organic farming. I will continue both conventional cultivation on leased-in land and organic cultivation on my own land.

Although organic farming was in general a traditional technology, organic *cotton* cultivation was a new technology for all the group members except the two that started organic cotton cultivation before joining the association. Although it is comparatively easy to show the economic benefit of Bt seeds, it is difficult to show the economic benefit of organic cotton unless it is purchased with a price premium. In the case of the association, Fairtrade certification did not work as a benefit that could compensate for loss incurred during the conversion period. The high market prices for conventional cotton also made organic farming less attractive, allowing farmers to buy expensive Bt seeds and other inputs.[15] After their three-year attempt at organic cotton cultivation, most of the members realized afresh that conventional farming with Bt seeds was the only option for profitable cotton cultivation.

3.5 A real confluence of Fair Trade and organic initiatives

This section first analyzes interactions between the two initiatives identified in the case study in two directions: (a) from organic to Fair Trade and (b) from Fair Trade to organic (see Subsection 3.2.2), and then discusses the consequent impact the confluence of the two initiatives has on the nature of the organic concept.

3.5.1 Interactions between the two initiatives

In the case of this farmers' association, the two initiatives look compatible in terms of direction (a). The attempt to pursue an organic certification began before the Fairtrade certification was given to the association. Although the association had not been certified as organic, the fact that the majority of members started organic cultivation must have contributed to the acquisition of Fairtrade certification. The reality observed in this case study corresponds to the first direction of theoretical interactions.

In the opposite direction (b), the Fair Trade initiative contributed to the promotion of organic cotton cultivation through Fairtrade premiums. Although not all the members participated in the sales to the Fairtrade market in the 2008–2009 season, Fairtrade premiums were shared equally by all. In all eight clusters Fairtrade premiums were, first of all, used as interest-free loans for purchasing non-Bt seeds for organic cultivation. Payment for production inputs at the beginning of each season is usually a financial burden on small and marginal farmers, which causes their indebtedness. Therefore, member farmers welcomed the cashless purchase of cotton seeds through AOFG and repayment of the loan without interest after harvest. This scheme facilitated member farmers to start organic cotton cultivation with non-Bt seeds distributed from the NGO. If all members can repay the loan regularly, the same money can be used as a revolving fund for purchasing seeds every year. The use of the remaining Fairtrade premiums was discussed and decided by the respective groups: some groups started the making of vermicompost (organic fertilizer including worm manure) collaboratively, one group purchased a pipeline for drinking water, and another made a drainage facility. The three groups in K village purchased six sprayers so that the group members could use them alternately when spraying organic liquid pesticides. Overall, it appears that the Fairtrade premiums were used for the promotion of organic practices.

At the same time, however, the Fairtrade premiums may also have contributed to the promotion of conventional farming with Bt seeds. As mentioned previously, the member farmers did not have to save money for purchasing non-Bt seeds at the start of a new season. Ironically, this tentative financial surplus enabled the members to purchase Bt seeds in cash. In the local shops there were two different prices for a single packet of Bt seeds: for instance, Rs. 700 in cash and Rs. 750 on credit. Poor farmers tended to buy seeds on credit, which increased their indebtedness. Most of the member farmers planting both Bt and non-Bt seeds purchased Bt seeds from a local shop in cash and non-Bt seeds from the NGO on credit without interest. In other words, the interest-free loans for non-Bt seeds enabled the members to get Bt seeds more cheaply than before. Furthermore, some of the members confessed that they used the sprayers that had been purchased originally for organic liquid pesticides for spraying chemical pesticides. It is undeniable that the Fairtrade premiums indirectly helped conventional farming with Bt seeds as well.

Unfortunately, the association members' sales to the Fairtrade market did not work as an incentive to continue organic farming during the conversion period. If the local market price had been low, the members might have realized the benefit of the minimum farm-gate price guaranteed by Fairtrade certification. Fortunately or unfortunately, however, the local market prices were constantly higher than the minimum price after the members joined the association. For them, selling to the Fairtrade market was no more advantageous than selling to conventional local traders. Arguably, Fairtrade certification by itself cannot support the conversion period of pre-organic farmers. As the experience of organic cotton projects in sub-Saharan Africa shows (Dowd, 2008), a strong price incentive, in addition to Fairtrade certification, will be required in order to keep attracting farmers to organic practices.

Regarding the second way of theoretical interaction in direction (b), what was the demonstration effect of Fairtrade-certified farmer groups? In the beginning, non-member farmers in the same village were very interested to see how the members would benefit from participation in the association, and especially how organic cotton cultivation would improve the livelihood of the members. Most of the non-members whom the author interviewed confessed that they had intended to start organic cotton cultivation if the members had been successful in keeping the same level of yields and obtaining price premiums. In reality, non-members judged the association's attempt unsuccessful not because of the lower yields of organic cotton but because of the cotton's organic nature that was not valued in monetary terms during the transitional period under the Fair Trade initiative. The non-member farmers also reached a conclusion that organic cotton cultivation was not a good choice. Unexpectedly, the confluence of Fair Trade and organic initiatives may have offered a majority of farmers in K village, both members and non-members, an opportunity to confirm the comparative advantage of Bt cotton rather than to perceive the value of the two initiatives.

On the whole, the interactions between the two initiatives did not lead to the promotion of organic cotton cultivation. Although this result obviously conflicts with the organic standard, it is not necessarily against the Fairtrade standard.

3.5.2 Impacts of the confluence

As this case study suggests, although both initiatives are opposed to GM organisms, the confluence of the two initiatives could paradoxically have prompted the spread of Bt cotton. This seems to be related to a clear difference between the two in terms of their primary objective: the organic initiative gives priority to how farmers *grow* crops and Fair Trade gives priority to how farmers *gain* from certified crops. In other words, the organic initiative puts less emphasis on how farmers gain during the conversion period; Fair Trade pays less attention to how farmers live their lives outside the initiative. This trend seems to be getting more and more distinct as the standards for the two initiatives weaken (see Subsections 3.2.1 and 3.2.2). If two certifications are obtained at the same time and put into practice in parallel, the two initiatives may be able to complement each other's weak points to some extent. In reality, however, there is usually a time difference in the acquisition of the two certifications: for small farmers, obtaining group organic certification is a longer process than obtaining Fairtrade certification is, as shown in this case study. Such a time difference will make the positive interactions between the two initiatives inefficient or unworkable. A likely outcome would be a failure in terms of complete conversion to organics or a mere pretense of the Fair Trade initiative.

The final question is whether the confluence with the Fair Trade initiative promotes the organic concept as "an alternative accumulation strategy for agrarian capitalism" or helps to retrieve its original nature as "an alternative framework for working the rural environment." Northern buyers' idea of double certification can be interpreted as a way of reinforcing the nature of both organics and Fair Trade as an alternative means of capital accumulation on the part of Northern buyers.[16] From the perspective of Southern producers, organic farming combined with a Fair Trade certification could become not a conventionalized concept "conducive to larger numbers of firms entering the [organic] market place" (Allen and Kovach, 2000, p. 224) but a new concept for helping marginal and small farmers. This new concept means in part "an alternative framework for working the rural environment" and partly releasing disadvantaged farmers from the existing agrarian capitalism that imposes financial burdens on them. This was, at least, the intention of AOFG when the NGO started the organization of small farmers.

Unfortunately, the new concept of organic farming was not realized in the observed case of the cotton sector in India. In this context, Bt cotton cultivation was more conventionalized than was organic cotton cultivation. The case study shows that the use of Bt seeds, appealing not only to large farmers but also to small farmers, increasingly pushes both into agribusiness players' "alternative accumulation strategy for agrarian capitalism." The "alternative accumulation strategy" is more powerful than the confluence of Fair Trade and organic initiatives. Given the lowering of standards for both initiatives, the confluence is likely to strengthen the nature of Bt seeds as an alternative means of capital accumulation, diluting organics' use as "an alternative accumulation strategy for agrarian capitalism" and "an alternative framework for working the rural environment."

3.6 Concluding remarks

This chapter has explored the confluence of Fair Trade and organic initiatives in the context of the prevalent Bt cotton production in India, applying Guthman's conventionalization thesis as an interpretative framework to the Southern context. In the case study in Telangana, India, the confluence of the two initiatives, contrary to their stated principles, contributed to the spread of GM seeds. Although the high market price trend for conventional cotton contributed to this paradox, it is also caused by factors embedded in the Fair Trade initiative. Fair Trade does not have a scheme for compensating for the decreased income that producers have to endure during the conversion period. Instead, Fairtrade premiums were used for the intrusion of GM seeds, reflecting Fair Trade's laxer attitude toward GM traits (see Section 3.1 in this chapter). Fair Trade's dilemma between helping poor farmers and promoting organic farming may have indirectly allowed the Fairtrade-certified farmer groups to incline to conventional farming with Bt seeds. As a result, the confluence of the two initiatives did not intensify the organic concept as "an alternative accumulation strategy for agrarian capitalism," but it did not release disadvantaged Southern farmers from existing agrarian capitalism, either. Rather, the confluence, contrary to its expectation, led farmers into another form of agrarian capitalism. Although the scope of this research does not enable us to judge which option is more beneficial to small cotton farmers – conventional cultivation with Bt seeds or organic cultivation – the outcome is obviously different from the original purpose of organic or Fair Trade, or both.

Finally, can we say that Fair Trade–organic double certification is better than a single certification? If the farmers' association in this case had pursued organic certification only, many of the member farmers might have chosen to continue organic practices in order to decrease production costs. However, it is not easy for small farmers to obtain and maintain organic certification and to compete with larger farmers in organic markets. If the association had pursued Fairtrade certification only, Fairtrade on its own might not have attracted members' interest because the Fairtrade certification itself guarantees neither stable markets nor better prices. For future organic price premiums, the members accepted enrollment in the association and compliance with all the Fairtrade standards. The confluence of the two initiatives seems to be a necessary strategy not only for Northern buyers but also for disadvantaged Southern producers. Through the confluence, however, the difference between the two movements appears more clearly. Whereas the organic is concerned with how farmers operate their land, Fair Trade is concerned with how farmers gain from the target crop. There is a gap between farming practice and income generation. It may be this gap that allowed Bt seeds to intrude in the livelihood of small farmers. The only way to fill the gap is to provide converting farmers with financial support such as pre-organic price premiums during the conversion period if organic cash crop production is used as a means of poverty alleviation.

The unsuccessful experience of this project in conversion into organics corresponds to the overall trend of India. A major decrease of organic agricultural land and producers occurred in India in 2010: while organic cultivated acreage increased from 1.02 million hectares in 2008 to 1.18 million hectares in 2009, the organic area shrank to 0.78 million hectares in 2010 (Wai, 2012, p. 170). It is reported that a large part of the reduction was "due to spread of Bt cotton and non-availability of non-Bt seeds" (Wai, 2012, p. 170).

Notes

1 Woods (2011) himself uses these terms regarding Guthman's conventionalization thesis in the Northern context.

2 Although organic certification prohibits the use of Bt seeds (IFOAM, 2007), Roy et al. (2007) and Roy (2010) report that some non-certified organic farmers in Gujarat, western India consider Bt cotton to be compatible with organic farming.

3 In 2014, the state of Andhra Pradesh was divided into two: Telangana and Andhra Pradesh.

4 In India the planting period takes place from March to September, and harvests from October to February (Osakwe, 2009).

5 The fourth type (d) includes natural farming with Bt or non-Bt seeds and organic farming in transitional periods.

6 When the project was implemented, the site belonged to the former state of Andhra Pradesh.

7 Plans submitted by groups in the other two villages had not yet been approved by the association.

8 *Mandal* is the administrative unit below the district unit. "Scheduled castes have been at the lowest end of the Hindu social caste hierarchy based on birth"; the social system-ascribed occupations of this group are generally "characterized by very low productivity." Scheduled tribes have been socially and economically underdeveloped because of "their long-time habitation in geographically isolated areas"; "lack of exposure to education and isolation from the social mainstream made them vulnerable to exploitation by non-tribals" (Sundaram and Tendulkar, 2003, p. 5263).

9 A one-hour interview with each respondent was repeated at least twice, although some respondents had more meetings with the author than others did. All interviews were conducted with the assistance of an interpreter.

10 This seems to be caused by a lack of proper information rather than by farmers' low education level. Exceptionally, however, a non-member farmer with 12 years of education had experimentally planted a variety of seeds on his own field.

11 Otherwise, farmers had to take a bus to a large seed market 26 kilometers away from the village.

12 Non-members who obtained non-Bt seeds through members planted both Bt and non-Bt seeds together.
13 One standardized packet of hybrid seeds, Bt or non-Bt, is used for one acre.
14 "Cultivating nontoxic . . . crops (refuges) in the proximity to transgenic crops that produce Bt toxins is widely recommended to delay pest adaptation to these toxins" (Vacher, Bourguet, Rousset, Chevillon, and Hochberg, 2004, p. 913).
15 Truscott, Lizarraga, Nagarajan, Tovignan, and Denes (2011) also note better prices for conventional cotton in India, which continued at least until June 2011, tempting contract growers to abandon organic cotton.
16 The concept of "Northern buyers" may include some Southern exporters and middlemen linking Southern producers with Northern buyers.

4 Plantation management and workers

A case from Darjeeling, India

As clarified in Chapter 1, a salient characteristic of Fairtrade certification in Asia is the large share occupied by the plantation sector. In contrast to a wealth of studies on Fairtrade-certified small farmer cooperatives, few studies were concerned with plantations that employ permanent workers in Asia as well as in other parts of the global South.[1] This chapter focuses on tea plantations, which significantly shape the reality of Fair Trade in Asia.

4.1 Fairtrade certification for the plantation sector

In the plantation sector, Fair Trade labeling has been mainly confined to a few crops such as bananas and tea for which small farmer cooperatives are underdeveloped (Jaffee, 2007, p. 217; Renard and Perez-Grovas, 2007, p. 150). The Fair Trade initiative was originally designed not for planters and plantation managers but for plantation workers. However, although owners and managers, like individual Fairtrade farmers, can benefit from the certification in the shape of access to a specialized export market and stable prices, wage-fixed workers are expected to benefit through (a) better working conditions that the Fair Trade label enforces socially and environmentally and (b) Fairtrade premiums used for worker communities (FLO, 2009; Jaffee, 2007, p. 218).

The indirect nature of these benefits for plantation workers creates two contrasting views on Fairtrade certification for the plantation sector. The first is grounded in the very reason for the introduction of Fair Trade to the sector. Supporters of plantation certification regard the Fair Trade label as "a powerful tool for forcing [planters] to improve labor conditions in industries notorious for abusing workers and exposing them to highly toxic pesticides" (Jaffee, 2007, p. 218). This view stresses the significant benefit that plantation workers gain indirectly from Fair Trade. The second view is a strong criticism that the primary beneficiaries of Fairtrade certification will be the plantation owners, not the workers (Besky, 2014; Renard and Perez-Grovas, 2007, p. 150). This view, by contrast, interprets the intangible nature of the benefits for workers as a negative impact of Fair Trade. However, both views, supporting and critical, fail to reflect the unique situation in which workers live and work. While Fairtrade certification can overtly benefit the management of a certified plantation by improving accessibility to markets and generating stable, higher prices, the benefits for workers depend on how the management offer them an enabling environment. Empirical research is needed to find out under what environment and by what method workers actually perceive the Fair Trade initiative.

In interpreting workers' perceptions, this chapter pays particular attention to the patron–client relations that have traditionally been established between the management and workers, especially in the tea plantation sector. The existence of the patron–client relations reveals a variety of tensions inside the domain of single certification. By adopting the perspective of patron–client relations as a form of power relations unique to tea plantations, we attempt to offer a third view on the Fair Trade initiative in plantations.

How do the existing patron–client relations influence workers' perceptions of Fair Trade? And in consequence, how can workers become beneficiaries or non-beneficiaries from Fairtrade certification under these circumstances? To explore these questions, this chapter draws on field research in a tea plantation in the Darjeeling district of India. The next section outlines the basic nature of the patron–client relations and their historical evolution in Indian tea plantations, and is followed by a methodological section and a description of the general patron–client relations in the plantation. Analyzing findings from this research, Section 4.5 clarifies plantation workers' perceptions of Fairtrade certification and the reasons for these perceptions. In the following sections we compare the perceptions of Fairtrade certification with those of two other interventions, namely, organic certification and a community development project for the plantation workers. Based on these analyses, the final section theorizes about the role of Fairtrade certification in the tea plantation sector and draws implications for its better use for plantation workers.

4.2 Tensions between management and workers in tea plantations in India

4.2.1 Patron–client relations

On the basis of the extensive anthropological literature, Scott (1972a, p. 92) defined the patron–client relationship as

> a special case of dyadic (two-person) ties involving a largely instrumental friendship in which an individual of higher socioeconomic status (patron) uses his own influence and resources to provide protection or benefits, or both, for a person of lower status (client) who, for his part, reciprocates by offering general support and assistance, including personal services, to the patron.

As Carney (1989) noted, this reciprocal relationship is actually expressed as *compliance*. The patron grants favors to gain compliance from the client in matters crucial to patronal interests; the patron must also comply, to some extent, with client preferences and needs. "The higher the degree of compliance, the stronger the relationship" (Carney, 1989, p. 45).

There are variations in patron–client ties. A level of client compliance can be coerced at different levels. Scott (1972a, p. 100), for instance, described different degrees of coercion as follows:

> In general, the oppression of the client is greater when the patron's services are vital, when he exercises a monopoly over their distribution, and when he has little need for clients himself. The freedom of the client is enhanced most when there are many patrons whose services are not vital and who compete with one another to assemble a large clientele.

A more oppressive or coercive form of relations does not necessarily mean better compliance. Likewise, the same level of compliance can be practiced with different levels of affinity to the client or loyalty to the patron (Lemarchand and Legg, 1972, p. 151; Scott, 1972a, p. 99). In this chapter, the strength or weakness of this relationship is discussed for the degree of compliance, not the degrees of affectivity and coercion.

The next two subsections consider how such patron–client relations have evolved between the management and workers in tea plantations in India.

4.2.2 *The origin*

India is the largest producer and consumer of tea in the world, and its tea industry rests on plantation-type production (Chakrabarti and Sarkar, 2005, p. 2). Tea plantations in India are a legacy of the British colonial administration. The colonial government in Assam, northeastern India, started the first tea plantation on an experimental basis in the 1830s, and within two decades private entrepreneurs developed and expanded tea plantations in northeastern India as well as in other parts of colonial India (Sharma and Das, 2009, p. 16; Singh, Narain, and Kumar, 2006, p. 2).

In the initial period of plantations, workers were recruited from different backward tribal areas, many of which were afflicted by famine (Behal and Mohapatra, 1992; Singh et al., 2006, p. 14). The objective of such a form of worker recruitment was to "retain a captive labor force at low wages" without developing a free labor market (Chakrabarti and Sarkar, 2005, p. 3; Siddique, 1995, pp. 91–92). Family rather than single settlement was encouraged, not only to prevent male workers from becoming recalcitrant or absconding but also to assure the reproduction of a labor force (Chatterjee, 2001, pp. 80–82). Females were also preferred as workers because of their sensitive fingertips, suitable for tea plucking, and their lower inclination to unionize (Sharma and Das, 2009, pp. 69–70). Once such migrant laborers reached the plantations, they were bound by contract and other coercive measures to serve for specific periods (Behal, 2010; Chakrabarti and Sarkar, 2005, p. 3). Planters were given powers to arrest laborers who absconded without warrant and to imprison those who refused to work (Behal and Mohapatra, 1992, p. 146). Thus, plantation work was sometimes described as "unfree" or "bonded" labor (see, e.g., Dasgupta, 1992).[2]

The plantation embracing immigrant workers was a kind of enclave in isolated regions. In these enclaves "living and recreation were integrated in a single system of control. Not only did laborers work in this system, but they were also born, grew up and married within the system; they had children here, brought up their families and died" (Alawattage and Wickramasinghe, 2009, p. 710). Nobody, not even policemen, could enter this kingdom without permission from the planter or managers. According to a report written in 1919, even if a manager assaulted or insulted a laborer or took girls from laborer families as his mistresses, there was no recourse to dispute the manager's actions or authority (Singh et al., 2006, p. 16). During British rule the planters were above the law; they treated their laborers inhumanely and ruled them tyrannically, considering them "uncivilized black barbarians" (Singh et al., 2006, p. 67). Until India's independence in 1947, plantation workers were not allowed to unionize themselves; there was no mechanism for collective bargaining (Chakrabarti and Sarkar, 2005, p. 4). In brief, the relationship between the management and workers during the colonial period can be described as a coercive "master–servant" relationship (Alawattage and Wickramasinghe, 2009).

Even during the colonial period tea estate workers were forced to depend on secondary incomes to supplement their plantation wages. Plantation workers were paid "below-subsistence" wages; earnings from plantation wage labor amounted to only 60 percent to 80 percent of family expenditure, resulting in high mortality and low fertility (Behal and Mohapatra, 1992; Dasgupta, 1992, p. 184). Such low wages, albeit usually leading to a lower degree of client compliance, tended to strengthen patron–client relations, complemented by another means of patronage. Planters gave workers plots of land to cultivate paddy and vegetables and allowed them to get involved in a variety of other activities, including "rearing and grazing cattle, raising poultry, collecting forest produce, wood cutting [and] fishing . . . within the garden grant and in its periphery" (Dasgupta, 1992, pp. 184–185). Although planters feared that allotting garden plots to workers would keep them away from tea garden work, this means of patronage functioned "as a new instrument for control" (Dasgupta, 1992, p. 184). Those allotted garden plots were bonded to the estate (Siddique, 1995, p. 107).

The arrival of Indian planters began before independence. In northeastern India, local elites – Bengali upper-class lawyers and well-to-do Muslim immigrants – began their plantation ventures almost as soon as the British did. These local elites derived from *zamindars*, or overlords appointed as intermediaries for land-revenue collection between the British authorities and the actual tillers of the soil (Jannuzi and Peach, 1980). "These 'native' planters shrewdly expanded their small plantation holdings and created styles of management and rule that were close to the feudal norms or pre/colonial Bengali and Assamese *zamindari* (landowning) cultures" (Chatterjee, 2001, p. 98). After independence, departing British planters were also replaced by a new cadre of Indian management, which took the form of small individual family-owned plantations, large Indian-owned corporations, or multinational corporations (Chatterjee, 2001, pp. 108–109). Chatterjee (2001, p. 155) described the patronage of postcolonial Indian planters as "the combination of British ideas of gentries lordship and indigenous ideas of *zamindari* entitlement."

4.2.3 Changing relations

Over time, the nature of the patron–client relations might have shifted from a more repressive to a more patrimonial one.[3] A landmark in the welfare of plantation workers was the Plantation Labor Act (PLA), enacted in 1951. This prohibits employers from inhumane behavior toward the employees of the plantations and exhaustively regulates the working and living conditions of workers, including their entitlements in housing, sanitation, drinking water, medical facilities, canteens, crèches, and primary schools (Sharma and Das, 2009, pp. 45–52). The PLA and other related legislation officially reformed the relationships between a planter and workers from master–servant to employer–employee. Even in the 1990s, however, there were very few tea plantations where all the provisions of the PLA had been fully implemented (Sharma and Das, 2009, p. 45; Singh et al., 2006, p. 59). Given this fact, the safe working conditions required by Fairtrade certification are by no means redundant, but remain in favor of the workers.

A characteristic of the relationships between management and workers is a distinct hierarchy in the plantation system, that is, the class structure of employees. As Chakrabarti and Sarkar (2005, p. 6) classified it, the vertical hierarchy broadly comprises four categories: (a) management, including plantation managers, assistant managers, and factory managers; (b) staff, including all white-collar employees in senior supervisory positions; (c) sub-staff, including all other subordinate employees in supervisory positions; and (d) rank-and-file workers engaged in either the field or factory operations. In the typical management style of plantations, orders issued from the top are relayed through this system, which serves to maintain and widen social distance between the management and workers as well as between the upper and lower rungs (Figure 4.1). Although trade-union activity in the plantation region has been remarkable since independence, union leaders are situated in the top tier of the postcolonial labor hierarchy and are granted "behind-the-scenes favors" by planters (Chatterjee, 2001, pp. 142–147). Such political alliances between planters and union leaders necessarily reinforce the hierarchy and social distinction in the tea plantations.

Postcolonial planters and managers have played the role of patron for plantation workers, properly using both the carrot and the stick. However, in recent decades patronage has become weaker in tea plantations, in line with a general trend observed between landowning employers and landless laborers in Indian agrarian society (see, e.g., Breman, 2000). To explain the weakening patron–client relations between management and workers, Sharma and Das (2009) pointed to the impact of globalization on the tea sector. First, increasing competition with other tea-producing countries requires plantation owners to produce

Figure 4.1 Female fieldworkers plucking tea leaves and a male supervisor (in the white shirt) inside the study tea estate (Darjeeling, India)

high-quality tea at low cost. To the owners, uneducated plantation workers have become unproductive human stock; planters started to curtail welfare expenses and reduce the number of permanent workers, instead paying low wages to casual laborers on a piece-rate basis. Second, tea plantation owners started to invest money earned from the tea industry in other economic sectors in response to the government's liberalization policy. Many tea estates were closed. These circumstances have compelled or encouraged plantation workers, especially males, to find employment outside the tea plantation sector. A survey conducted in the Darjeeling district in 2007 revealed that 69 percent of the family members of sample worker households worked outside the plantation sector (Sharma and Das, 2009, p. 92). This consequence of weakened patronage from plantation management implies emancipation from the traditional bondage system as well as increased uncertainty and vulnerability in workers' livelihoods.

4.3 Methodology

4.3.1 Patron–client relations as an analytical framework

Fairtrade certification was introduced to tea plantations to take advantage of globalization. Inevitably it has interacted with today's weakened patron–client relations between management and workers. The influence of patron–client relations on workers' perceptions of Fairtrade certification is analyzed from two aspects: (a) what influences the existing power relations have on the shaping of workers' perceptions, and (b) whether certification helps strengthen or further accelerates the weakened relations. In other words, workers' perceptions are interpreted as the consequence of the long-established power relations on the one hand, and the impact of Fairtrade certification on the relations on the other.

Although patron–client relationships between the management and *male* workers must, in reality, be different from those between the management and *female* workers, the scope of this research does not extend to shedding light on gender differences. This chapter deals with worker households rather than individual workers as the analytical unit, and focuses on matters common to both male and female workers.

4.3.2 A tea plantation as the unit of case study

The Darjeeling district in West Bengal is one of the major tea-producing areas in the world. Of its 87 tea plantations, about 50 percent are certified as organic (or in the process of conversion), and 30 percent are Fairtrade.[4] While most Fairtrade-certified estates belong to large-scale companies, there are a few family-owned independent plantations certified as both organic and Fairtrade. One of the latter, called the Sonapahar tea estate (pseudonym), was selected as a case study for two reasons. First, it was the first plantation in the Darjeeling district to acquire both organic and Fairtrade certifications. When the first author, Makita, did her fieldwork in May–June 2009, the plantation had already experienced 16 years of organic export and 15 years of Fairtrade export. Fairtrade certification was therefore expected to have permeated the estate workers more fully here than it would in the other certified tea estates. Second, the managerial system of this family-owned independent plantation seemed to be more suitable for the direct observation of patron–client relations between management and workers than were other complicated styles of management seen in conglomerates.

The history of the Sonapahar tea estate began when a British officer bequeathed the official registration of his tea plot to a young lad from an affluent Bengali *zamindari* family in 1859. The estate was owned by the fourth generation of the founder in 2009. Unlike other estate owners, who preferred to live in Kolkata (the capital of West Bengal), the owner of Sonapahar lived on the estate. Therefore, more correctly, this plantation had five rungs of hierarchy, with the owner above the management. In May 2009, 555 permanent workers of Nepali origin worked and lived with their families on 670 hectares of the hilly land spreading from 550 to 1,260 meters above sea level. The estate included seven worker settlement villages divided into nine communities. All the produce from Sonapahar was 100 percent organic, and 60 percent to 80 percent of the export, varying from year to year, went to the Fairtrade market.

In recent years a non-governmental organization called Community Health Advancement Initiative (CHAI) had assisted worker communities in nine tea plantations, including Sonapahar and another two Fairtrade-certified estates in the Darjeeling district, in the fields of health care, small-scale rural infrastructure, income generation, and the organization of adolescents (as of June 2009). In Sonapahar the CHAI project started in 2005. The owner of the Sonapahar estate agreed to spend part of the Fairtrade premiums on the CHAI project on the estate.

4.3.3 Workers' perceptions as data

Observing how plantation workers perceive the Fair Trade initiative is central to the case study. Workers' perceptions of the two other interventions linked with Fairtrade certification, namely, organic certification and the NGO's community development project, are also analyzed. Organic certification is linked with Fairtrade certification in two ways. First, because a requirement of Fairtrade certification is to comply with organic or sustainable production practices, organic certification facilitates the acquisition and maintenance of Fairtrade certification. The Sonapahar estate itself acquired Fairtrade certification in the wake of organic certification. Second, the International Federation of Organic Agriculture Movements (IFOAM) aims to relate to "workers' rights, their basic needs, adequate return and satisfaction from their work and a safe working environment" (Browne et al., 2000, p. 83), which is resonant with Fair Trade guidelines. Although plantation workers gain no monetary forms of benefit from organic certification, it can reinforce the first benefit expected from Fairtrade certification, namely, better working conditions. In contrast, community development implemented through the CHAI project corresponds to the second benefit of Fairtrade certification, namely, Fairtrade premiums being used for workers' communities. By identifying similarities and differences among the perceptions of the three interventions, this chapter attempts to theorize about a plausible way for plantation workers to take advantage of Fairtrade certification.

4.3.4 Data collection

Data collection focused on worker households living in one community that was close to both the factory compound and the highway connecting to two major cities in the Darjeeling district. The residents in this community were therefore expected to have access to more information than those in remote villages did. Primary data were collected from all 62 households in the community on how the workers perceived the benefits of the Fair Trade initiative, organic cultivation, and the CHAI project through semi-structured interviews. The workers interviewed might have been more familiar with Fairtrade and organic activities than were workers in the other communities, implying that if this focus group did not know, neither did average workers.

To interpret the perceptions of the workers and to confirm information from the focus community, supplementary interviews were undertaken with non-worker family members of the 62 households; the management, staff, and sub-staff of the plantation; some workers living in the other eight communities, including trade union members; and the staff of CHAI.

4.4 The plantation and the focus community under patron–client relations

There were typical patron–client relationships in the Sonapahar estate. Although it is impossible for an outsider to observe all the elements of reciprocity between patrons and clients, the author noticed all five patron-to-client flows identified by Scott (1972b, p. 9), namely, (a) basic means of subsistence, (b) subsistence crisis insurance, (c) protection, (d) brokerage and influence, and (e) collective patron services. The management provided all estate workers with steady employment, land for living, and other basic services in compliance with the PLA. Even in years of poor harvests it guaranteed the same levels of wages and services by absorbing losses. In addition, it gave loans or grants in the case of sickness or accident in workers' families. At the time of natural disasters the management negotiated with the local government and outside supporters for the repair of workers' houses and common facilities. It is also possible to understand Fairtrade certification as an example of the "brokerage" function; the owner of Sonapahar used his power and influence, at least partly, "to extract rewards from the outside for the benefit of his clients" (Scott, 1972b, p. 9) by obtaining and maintaining Fairtrade certification. Finally, in cooperation with the owners and managers of other estates he promoted organic cultivation in the entire Darjeeling tea sector; such promotional activities led to the economic security of workers in Sonapahar as well as in other estates in the district. By contrast, the flows of goods and services from clients to patrons were, as Scott (1972b, p. 9) conceded, "hard to characterize since, as his patron's 'man,' a client generally [lent] his labor and talents to the patron's designs, whatever they may be." Estate workers, both permanent and temporary, were employed at the same wage rate not only for tea cultivation but also for personal domestic services in the houses of the owner and managers. The patron–client relationship in Sonapahar was symbolized by the two names the workers used to describe the owner of Sonapahar: the king of the estate and the father of the workers.

In agreement with the findings of Sharma and Das (2009, p. 92), the livelihood of Sonapahar workers was heterogeneous. As Table 4.1 shows, 55 of the 62 households in the focus community contained at least one permanent estate

Table 4.1 Surveyed households by number of estate workers (Darjeeling, India)

No. of estate workers	*No. of households*
1 permanent	20
2 permanent	17
3 permanent	8
1 permanent + 1 temporary	7
2 permanent + 1 temporary	3
None*	7
Total	62

Source: Field survey

Note: *Includes a retired worker.

worker. In the remaining 7 households (11 percent), none of the family members worked on the estate as a permanent or temporary worker; nevertheless, they were allowed to continue to live inside Sonapahar because their deceased parents or grandparents had been permanent workers. These non-worker families, who were provided only with the free land space by the estate, were also regarded as members of the community and were automatically included in the beneficiaries, for example, if a Fairtrade premium was used for a community service or infrastructure. Most of the non-worker households had a stable job outside the estate and were proud of the better-looking houses they had built with their own savings.[5]

Even in the households that contained permanent workers, the main income did not necessarily come from the wages from the estate (Table 4.2). While the majority (44 households) relied primarily on wages from the estate, in 10 households non-estate income exceeded estate income. In each of these 10 families the husband worked outside the estate, keeping the wife's job on the estate. In reality, out of the 89 permanent workers in the 62 households, 52 (58 percent) were female fieldworkers mainly involved in tea plucking. Such male mobility is partly because of the *Badli* system introduced by British planters in the Darjeeling tea industry, under which only one child of a worker is employed after his or her retirement from plantation work (Sharma and Das, 2009, p. 81).[6] Since it is usually easier for men to find employment outside the tea plantation sector, the work right is, in many cases, taken over by a female family member. Under the *Badli* system, the management cut back the personnel and welfare expenses and seemed to allow, in return, non-workers and their families to continue to live on the estate free of charge. Between the management and those families that did not rely on estate work, there still existed a certain level of reciprocity, albeit lower than that observed between the management and those reliant on estate work.

Job seeking outside the tea plantation sector usually leads to two scenarios: (a) engagement in a stable job such as employment in a government office, the

Table 4.2 Surveyed households by main income source (Darjeeling, India)

Main income source	No. of households
Permanent estate work	42
Temporary estate work	2
Non-estate work:	16
Compost making	(1)
Security guard	(4)
Employment in government offices	(4)
Employment in private companies	(2)
Shop management	(2)
Army	(1)
Police	(1)
Taxi driver	(1)
Pension	2
Total	62

Source: Field survey

army, or a large private company, or (b) work as a temporary wage laborer inside or outside the tea estate. Those who achieve the former are better educated and blessed with connections with helpful people. While 35 (57 percent) of the family heads surveyed had completed only primary-level education (up to Class 5), all the family heads involved in non-estate work boasted an educational background of Class 8 or higher. However, it is not easy to find a stable job outside the tea estate, even with a good educational background. The author met many young males who had completed high school or college-level education but who had not found a job; the educated men would not accept employment as physical laborers but remained on the estate simply as dependents of their parents. As long as the younger generation depend on their parents, they can continue to live under the patronage of the estate management.

In summary, the increasing mobility to non-plantation work has generally weakened patron–client relations between management and workers on the one hand, and on the other hand has generated economic disparities among worker households that were equally poor when they or their ancestors arrived on the estate. At present, the clients can be divided into two groups: those who do and do not have to rely heavily on the patron.

4.5 Fairtrade certification under patron–client relations

4.5.1 Invisibility as a consequence of existing power relations

None of the people living in the focus community, apart from one office worker, knew the concepts of the Fair Trade initiative and premiums from the certification; the term "Joint Body" was more familiar to some people. The author therefore asked people in Sonapahar not about Fair Trade but about the Joint Body's activities. According to Fairtrade International (FLO), a Joint Body of the Fairtrade scheme was a group of *elected* worker representatives organized to use Fairtrade premiums democratically (FLO, 2009).[7] Although the latest Standard for Hired Labor revised in 2014 uses "Fairtrade Premium Committee" in place of Joint Body (FLO, 2015), in this chapter we use Joint Body, known in the case study venue. Table 4.3 shows the community people's knowledge of the Joint

Table 4.3 Surveyed households' perceptions of the Joint Body (Darjeeling, India)

Perception levels	No. of households
Never heard	11
Know only by name	18
Have some knowledge	33
Know some members only	(10)
Know some activities only	(9)
Know some members and activities	(14)
Total	62

Source: Field survey

Body. A more positive answer from each household was taken as the perception of the household: if at least one member had some knowledge, the household was regarded as having knowledge; only when none of the adult members of a household had ever heard the name was the household counted as never having heard of it.

Although about half of the households had some knowledge of the Joint Body, they knew nothing about its financial sources. They guessed that it was financed by the management or donations from foreign well-wishers. Although 24 families (39 percent) recognized a few members of the Joint Body, the author did not come across anyone either in the focus community or in the other worker communities who had actually participated in the selection process of Joint Body members. Those who knew some Joint Body members had merely had opportunities to see them attending a meeting in the head office. Although reports from the estate office list a variety of projects in which Fairtrade premiums were invested in recent years (Table 4.4), only microcredit, which was suspended in 2000, was significantly identified as an activity of the Joint Body (Table 4.5). These facts suggest that rank-and-file workers have barely been informed about Fairtrade premiums, which make up an extra budget for workers' welfare as well as a return from the sales of tea – the fruits of their hard work.

Despite such low recognition by workers, a Joint Body had existed in this plantation and Fairtrade premiums had been used to benefit workers and their communities for the past 15 years. The author met the people whom the focus

Table 4.4 Expenditures of Fairtrade premiums on the study tea estate (Darjeeling, India)

Categories	*April 2006– March 2007 (%)*	*April 2007– March 2008 (%)*
1. Home stay project	5	22
2. Plant nursery making	25	2
3. Tree planting	18	24
4. Compost making	10	6
5. Biogas project	9	11
6. Other domestic energy sources	0.1	–
7. Medical treatment	16	9
8. Education for workers' children	2	2
9. Social ceremonies	4	1
10. Food and transportation	1	7
11. Other welfare purposes	6	10
12. Salary for additionally generated employment	1	5
13. Expenses for visitors	5	–

Source: Reports from the tea estate studied

Note: Items 1 to 4 aim to generate additional income for worker households.

Table 4.5 Surveyed households' perceptions of benefits from Fairtrade premiums (Darjeeling, India)

Perceived benefit	No. of households
Not received anything	22
Received*:	40
Microcredit from the Joint Body	(15)
Microcredit from the estate	(19)
Cash grant from the Joint Body	(1)
Cash grant from the estate	(8)
Materials from the estate (for bio-gas)	(1)
Temporary employment from the estate (tree planting)	(1)
Training (medical) from the Joint Body	(1)
Total	62

Source: Field survey

Note: *Some families referred to more than one kind of benefit.

community recognized as Joint Body members. All came from the staff or sub-staff; no workers from the lowest category of the hierarchy were included in the Joint Body. Two of the current members confessed to the author:

- Female paramedic in the health center: "Invited by a neighbor, I happened to attend a meeting. It was a meeting of the Joint Body. Then, I was also appointed as a member on the spot."
- Male chief field supervisor: "I was asked by the previous chief manager [who had died] to attend a meeting of the Joint Body only once. I am just a reserve member, not an official member."

The current Joint Body members were most probably not elected by workers but appointed by the management, although the owner and managers, in conversations with the author, insisted that the current members were representatives of the workers.

There were reasons for the invisibility of the Fair Trade initiative among the workers. The current chief manager conceded that it was actually difficult for uneducated workers to understand the system of Fairtrade premiums and to manage the fund. He also mentioned that some active trade-union members wanted to take advantage of the Joint Body for their radical activities.[8] He recollected:

Some trade-union members asked me to choose them as members of the Joint Body. We [the management] refused their offer because the Joint Body should be a non-political organization. Getting angry, they instigated workers who had used loans from Fairtrade premiums to default on the repayments.

The management feared that the trade union might empower itself by organizing workers under the name of the Joint Body and by funding its anti-management activities with Fairtrade premiums. Complaints to the current Joint Body were confirmed in the author's interview with a trade-union member, who told her about the union's recent meeting with the management: "We have not been able to see what the Joint Body does. We therefore proposed that the management should change the current Joint Body members." Because the management carried the registration and inspection fees for the Fairtrade label, it did not want to use the premiums for activities against it. It was natural for the management to avoid informing workers about the Fair Trade initiative in order to maintain the hierarchical system on the estate.

As a consequence, the management had made a substantial number of decisions on how to use the premiums. This was endorsed by some of the Joint Body members:

- We do not have meetings regularly. We only do so when a visitor, such as a foreign volunteer, comes to the estate.
- I do not know who records the expenditure of the premiums. I have never seen such a record.
- I do not know why microcredit was suspended. The management did not explain the reason to us. None of us asked the reason at meetings, either.
- Workers submitted microcredit applications to the manager through one of the Joint Body members. The manager himself screened these applications from workers and decided the amount of each loan.

Such an inactive involvement of the members in the Joint Body was caused not only by the management's intentions but also by the members' reluctance to spend extra time voluntarily on Joint Body-related activities without monetary remuneration. The current members were indebted to the management in various ways: some had been given special training opportunities; some were promoted to good positions more quickly than others. Therefore, to maintain good personal relationships with the owner and managers the appointed members spared the minimum time and labor for duties such as attending meetings arranged for visitors and replying to questions from Fairtrade inspectors. The inertia of the Joint Body, favorable to both the management and the members, was maintained under their patron–client relationships. Consequently, the members did not have much information they, as worker representatives, could convey to the other workers.

4.5.2 *The impact of Fairtrade certification on existing relations*

Even without knowledge of the Joint Body, many worker households had enjoyed substantial benefit from Fairtrade premiums. In the community surveyed, while 22 households (36 percent) replied that they had not individually received any extra services or benefits other than those stipulated by the PLA, the other 40 households had received additional welfare assistance (see Table 4.5). All these forms of assistance were, in practice, funded from Fairtrade premiums, although many of the families believed that this assistance came from the management's generosity. Workers who had received special assistance from the management showed gratitude toward the owner and managers. In other words, the workers perceived that the degree of patron compliance had become higher. Although the author did not witness this greater degree of compliance on the part of the workers, they must have theoretically responded to the patron's new compliance level.

Why the current Joint Body members accepted extra unpaid duties imposed on them can also be explained by the tangible benefits from Fairtrade premiums. A female member confessed her reasons for continuing to be a member:

> For the sake of people in my community. When somebody in the community becomes sick, the management can bear the medical expense with the premiums.

In such a hierarchical society there was also a patron–client relationship between this member and the people in her community. By consenting to be a Joint Body member she could increase the patron-to-client flow from herself to the community people. At the same time, for her to consent to be a member was a kind of incremental client compliance toward the management. Priority might have been given to Joint Body members in the distribution of Fairtrade premiums.

Although the premiums were used not only for such assistance to individuals but also for the operation of community facilities such as a library and a crèche, the workers had few opportunities to get to know the facts. Because of the invisibility of the Fair Trade initiative on the estate, the premiums simply worked as an increase in the patronage of the management toward specific individuals.

4.6 Organic certification under patron–client relations

The Sonapahar tea estate had practiced not only organic agriculture but also the biodynamic farming methods advocated by Rudolf Steiner. Tea products from Sonapahar were being certified by Demeter (the certification body of biodynamic farming). Although the name of Demeter was conspicuously displayed inside the factory as well as facilities for making organic fertilizers, most workers passing by the indications every day did not know what they meant. Like Fairtrade certification, the concept of organic certification had not been explained to them. Workers never asked their superiors questions deemed to be unnecessary for their daily work. Therefore, how they understood organic cultivation, not certification, was central to the author's interviews with them.

Organic cultivation was more closely linked with workers' daily lives than the Fair Trade initiative. Regarding the 63 field-related permanent workers in the focus community, the author first asked whether they knew the difference between organic and chemical fertilizers and pesticides (Table 4.6). Of those interviewed, 65 percent did not understand the concept of organic cultivation. Most of the respondents had never seen chemical fertilizers because organic farming had been substantially practiced for the past 30 years following the current owner's succession

Table 4.6 Estate workers' perceptions of organic cultivation (Darjeeling, India)

Difference between organic and chemical	No. of fieldworkers
Have some knowledge	22
Do not know:	41
Nonetheless, Use cow dung and rotten leaves	(39)
Use vermicompost	(1)
Do not recognize any	(1)
Total	63

Biodynamic farming	No. of fieldworkers
Have some knowledge*	19
Do not know	44
Total	63

Source: Field survey

Note: *Most fieldworkers did not know the term biodynamic but referred to some materials used for biodynamic farming.

to the management of the estate from his father. The author then asked them what kinds of fertilizers they used. All but one worker quickly mentioned cow dung and rotten leaves. These were traditional organic manures that had always been close to their lives; they had used these manures not only in the tea gardens but also in their own small home gardens. Despite practicing organic farming, they were unaware of its value. The remaining 22 workers (35 percent) referred to certain differences between their ways of cultivation and the use of chemical fertilizers and pesticides. A typical answer was that organics were better for the health of people (including fieldworkers themselves) and the natural environment.

In brief, the workers' perceptions of organic farming were not particularly influenced by patron–client relations. Each worker's perception was contingent on his or her personal experience or beliefs. Some workers had learned the value of organic cultivation from their educated children.

Untraditional organic fertilizers, such as the recently introduced vermicompost and other peculiar materials used for biodynamic farming, were less familiar to workers (see Table 4.6). This might be because of the division of labor on the estate and because no opportunities were given to workers to ask questions. A female fieldworker mainly involved in tea plucking said:

> We [tea pluckers] sometimes apply fertilizers. Since we use fertilizers fetched by male fieldworkers, we know neither where the fertilizers came from nor who made them.

In fact, only appointed workers were involved in the key activities of organic farming, such as making biodynamic fertilizer and vermicompost. Eighteen years ago only one worker learned the biodynamic farming method from the estate owner. This worker was close to the owner (he was in charge of his personal horse). When the worker reached retirement age, a new temporary worker was selected to succeed to the job in biodynamic farming. This new worker was unemployed, taking casual care of the estate's cows and carrying milk to the owner's residence every morning. The newly appointed worker stated that the owner promised to make him a permanent worker if he mastered the biodynamic method.

The owner had also encouraged unemployed young men and temporary workers to operate compost-making facilities on the estate. The author witnessed that organic farming, which is more labor-intensive than conventional farming, was tactfully linked to the generation of additional employment opportunities for unemployed or underemployed estate residents. This additional employment generated by organic certification can be regarded as a rise in the compliance level on the part of the management, definitely leading to stronger patron–client relationships between the owner or management and the workers helped or benefited. Many of these additional employment opportunities were, in fact, funded from Fairtrade premiums (see Table 4.4), although none of the employed people had a means of knowing this.

4.7 The NGO's community development project under patron–client relations

In contrast to the invisibility of the Fair Trade initiative, the CHAI project was remarkably familiar to the worker communities (see Table 4.7). In each community within Sonapahar, the CHAI staff initially held a number of meetings with community members and helped them to discuss and identify what to do and how to do it in order to improve their lives. In the focus community the people decided to build latrines for those households without one, using materials provided by CHAI. Regardless of the fact that their period of participation was relatively short, from 2005 to 2008, the people of the community were impressed by their experience of participatory community development. This was because they had been accustomed to asking the management for help, but instead they were now able to solve their problems by themselves for the first time. In other community villages within Sonapahar people also proudly showed the author a variety of community infrastructures and facilities as the outcomes of their own projects.[9] The workers' perceptions of the CHAI project were influenced not by the existing power relations, but simply by their own participation in the activities.

Those who actively participated in the CHAI project did not know that a significant portion of the funding for the project came from Fairtrade premiums. The workers who benefited never recognized the CHAI project as an outcome of

Table 4.7 Surveyed households' perceptions of the CHAI project (Darjeeling, India)

Perception levels	No. of households
Never heard	4
Know only by name	4
Participated in the meetings but not benefited	18
Participated in the meetings and benefited	36
Total	62

Source: Field survey

the Fair Trade initiative because they did not know about Fair Trade itself. The invisibility of the Fair Trade initiative, which led the workers to regard cash grants and employment funded by the Fairtrade premiums as a sign of the generosity of their management, simply made them believe that the NGO project was a gift from the NGO. One difference was that, whereas the management-led welfare assistance resulted in the reinforcement of patron–client relations, the CHAI project did not influence existing power relations on the tea estate. The invisibility of the Fair Trade initiative on the estate hid the patronage of the management. As long as the real picture of the Fair Trade initiative remains hidden, the management are unable to reveal the fact that they decided to invest part of the premiums in the CHAI project.

Ironically, the invisibility of the Fair Trade initiative enabled workers to take action by themselves for the benefit of their own communities. The experiences of the CHAI project in the Sonapahar estate remind us that FLO sees Fairtrade certification in the plantation sector "as a means to increase the empowerment and well-being of workers" (FLO, 2009). It can be argued that the expectations of FLO may have been embodied by the workers who participated in the CHAI project if "empowerment" is interpreted as "the process by which people . . . who are powerless . . . develop the skills and capacity for gaining some reasonable control over their lives" (Rowlands, 1995, p. 103). Although "the skills and capacity" that the workers in Sonapahar acquired through the CHAI project are small, they were certainly the first step toward "the empowerment and well-being of workers" outside of the existing power relations.

It is true that this type of participatory community development could not be achieved without the intervention of the NGO. It was impossible for the Joint Body to play the role played by CHAI under the hierarchical system of plantations. This system had traditionally worked only for top-down, and never for bottom-up communication. The lowest category of workers were not allowed to ask questions of the staff or sub-staff who constituted the Joint Body; the staff were not able to ask questions of their managers; and even the managers were reserved in front of the owner. Only a third-party body such as CHAI was able to communicate properly with all the different strata of the hierarchy. The experience of CHAI cannot be generalized because, among the three Fairtrade-certified tea estates where CHAI operated, only Sonapahar agreed to use part of the premiums for the CHAI project. Nevertheless, we can learn the following lesson: the intervention of a third party in the operation of Fairtrade premiums opens up a promising opportunity to make the Fair Trade initiative substantially more visible in such a hierarchical society.

4.8 Making certification work for plantation workers

4.8.1 Comparisons of the three interventions

As Table 4.8 summarizes, the existing power relations had a significant influence on how estate workers perceived the Fair Trade initiative, but not on how

Table 4.8 Interactions of the patron–client relations with three interventions (Darjeeling, India)

	Influence of patron–client relations on workers' perceptions	*Impact of interventions on patron–client relations*
Fair Trade	Significant	Significant
Organic	Not significant	Significant
CHAI (NGO) project	Not significant	Not significant

they perceived organic cultivation and the NGO's community development project. While patron–client relations made the Fairtrade scheme completely invisible to the workers, organic cultivation was almost embedded in their everyday work and lives, irrespective of whether it was certified. Despite this difference in perception, both certifications contributed to the reinforcement of existing patron–client relations. Although organic certification itself did not create the contradiction observed in the case of Fairtrade certification, it helped to reveal the contradiction.

By contrast, the NGO's project was free from the influence of the existing power relations: the patron–client relations did not hinder the workers' perceptions of the project, and the project neither strengthened nor weakened the patron–client relations.

4.8.2 How to take advantage of the power relations

The patron–client relations strengthened by Fairtrade and organic certifications then enable the management to mobilize workers more easily for Fairtrade and organic tea production. This finding does not challenge the typical criticism of introducing a Fair Trade certification program into the plantation sector. However, the strengthened patron–client relations do not always mean a problem on the part of workers.

The patron–client relations between management and workers will be maintained as long as the owner and managers want to continue to operate the plantation and as long as workers want to continue to work there. In Sonapahar there are a cluster of people who can make a living without patronage from the estate management. Such better-off people have to maintain only a minimum level of compliance in exchange for their living on the estate's land. For those who cannot help but work on the tea estate, weakened patronage from the management means more vulnerability in their lives. Under the current plantation system, therefore, workers appreciate patronage reinforced by Fairtrade and organic certifications, which, although they do not have the power to change the traditional patron–client relations into equal employer–employee relations, contribute to the betterment of workers' lives in hierarchical plantation society.

One problem is that the increased patronage through the certifications is not equally distributed among the workers. Since the amount of Fairtrade premiums is limited, financial assistance does not reach all households; the number of jobs created in organic farming does not equal the number of unemployed people on the estate. The management naturally patronize certain specific workers only, the selection of whom depends on the management's arbitrary decisions. Consequently, such biased patronage can lead to discord in the worker communities.

The case of the CHAI project in the Sonapahar tea estate suggests a solution to this problem. Under the current Fairtrade scheme the premiums simply mean an increase in the welfare budget of the tea estate because most workers are not informed of the existence of the premiums. It is technically possible for the management to misuse the increase in the welfare budget through the current Joint Body system. Any form of internal monitoring would not work under the hierarchy. As the experience of the CHAI project suggests, the only way of using premiums equally and properly is to allow an independent body to operate as an intermediary. If a premium is used for a third party's project, it directly reaches worker communities without reinforcing patron–client relations. Even if they know nothing about the Fair Trade initiative, the concept can substantially empower plantation workers and improve their living conditions through an intermediary. The management of the Sonapahar tea estate eventually offered an enabling environment to its workers by investing Fairtrade premiums in the NGO's project.

FLO is encouraged to play a central role in the selection of such a third-party body and to contract administration of the premium to an intermediary organization as a prerequisite for certification. This single case study does not enable us to draw up the picture of an intermediary organization suitable for this role; more research will be necessary in actual plantation settings. It is, however, clear that such an intermediary organization should be independent of both management and trade unions. Although FLO (2009) expects Fairtrade certification to increase the empowerment of workers, the management of tea plantations will never accept Fairtrade certification if it motivates workers to demand a wage increase or the resignation of the management through collective action. The empowerment of workers has to be within the management's tolerance level.

4.9 Concluding remarks

This chapter started with a contradiction within Fairtrade certification for the plantation sector: while the certification is designed for plantation workers, benefits to workers are controlled by the plantation management. On the premise of this contradiction, impacts of Fairtrade certification were explored from the perspective of tea plantation workers, focusing on the certification's interactions with the patron–client relations between plantation management and workers. In agreement with Besky's (2014) findings, the present case study revealed that the Fair Trade initiative was barely known to the workers and that Fairtrade premiums were not managed by workers' representatives. However, it was also true that workers had unknowingly benefited from Fairtrade premiums. The invisibility of the Fair Trade initiative simply reinforced the existing patron–client relations. Organic certification also contributed directly and indirectly to this reinforcement through Fairtrade certification. In other words, double certification contributes more to the reinforcement of the existing power relations than does single certification. Given that the traditional patron–client relations have been weakened, this reinforcement may be favorable to those workers who are unable to find alternative employment outside the plantation sector.

This case study also revealed that the benefits of the certification scheme did not reach all workers in need of them equally. A solution to this problem may be to invest Fairtrade premiums in community development projects led by a third-party organization independent of the hierarchical society of tea plantations. In the present case, when the operation of the premiums was transferred from the plantation management to a third-party body, the invisibility of the Fair Trade initiative inversely hid the patronage of the management and instead contributed to the empowerment of workers. Despite the essential contradiction, Fairtrade certification can help plantation workers if transparency is guaranteed in the use of premiums.

Notes

1 Except for Besky (2014), which focuses on the tea plantation sector, and Ruben and van Schendel (2008), which focuses on the banana plantation sector, a few authors such as Renard and Perez-Grovas (2007) and Valkila and Nygren (2010) have referred to both independent producers and employed workers in the coffee sector.

2 The concept of unfree labor also included debt-bonded labor (Alawattage and Wickramasinghe, 2009).

3 Hall (1974, p. 507) referred to two types of patron–client relations: "those based on overt acceptance of traditional values by the subordinate (patrimonial) and those based increasingly on more obvious forms of repression by the powerful . . . (repressive)."

4 Based on information from the Darjeeling Tea Research and Development Centre (as of June 2009).

5 The management provided workers with uniformly rustic housing. Some better-off workers repaired, renovated, or extended their houses themselves.

6 The work right is also traded informally outside the *Badli* system.

7 FLO allows a minority of management representatives to join the Joint Body and to assist in the operation of the premium fund (FLO, 2009).

8 Stevis (2015, p. 103) views that the revised Standard "strongly encourages [workers] to move toward collective bargaining in a deliberate fashion." To observe the impact of this revision on relationships between the management and the trade union will be a future research topic.

9 In the other worker communities of Sonapahar the CHAI project resulted in village roads, footpaths, community centers, latrines, a playground, and a water supply system.

5 Social movements and commercial certification

A case from Thailand

This chapter attends to the contradiction between a movement and certification that has frequently been identified within previous studies. The transformation of Fair Trade and organic initiatives from social movements into certification-based export businesses has been observed in the case of organic rice production in northeastern Thailand. However, continued commitment to values originating in social movements, despite growing commercialism, has also been observed among some farmers' groups in Thailand.

5.1 Organic and Fair Trade initiatives linked to homegrown social movements in Thailand

Many scholars have critically examined the processes of commercialization and institutionalization of organic and Fair Trade initiatives in the global North. Ultimately, these processes have been shown to undermine the values of movements that originally pursued the lofty goals of transforming conventional agri-food systems. These original transformative visions of movements around organic agriculture are evidently on the wane vis-à-vis increasing commercialization of organic farming in Europe, the United States, and Australia (Guthman, 1998; Kaltoft, 1999; Lockie and Halpin, 2005; Tovey, 1997). A similar process can also be observed regarding the Fair Trade movement that has become more commercially oriented in tandem with the expansion of certification-based businesses (e.g., Lyon and Moberg, 2010; Raynolds et al., 2007).

Northern actors such as development NGOs, charitable organizations, and private traders have introduced Fair Trade and organic agriculture initiatives to the global South (Bolwig, Gibbon, and Jones, 2009; Lyon and Moberg, 2010; Raynolds, Murray, and Wilkinson, 2007). When these initiatives are driven by export incentives, discrepancies between the intentions of foreign buyers and local producers' concerns are likely outcomes. Thavat's (2011) study on organic rice production in Cambodia revealed a typical example of discordance between an externally imposed organic project and local farmers' needs. Thavat argued that the project's emphasis on growing new export-oriented varieties of aromatic rice, which were not always suited to local agro-ecosystems, eventually led to a reduction of overall paddy yields and higher costs regarding labor and inputs, thereby creating an adverse impact on farmers' lives. Similar discrepancies have been observed between organic and Fairtrade certification schemes and local socioeconomic contexts within rural Latin America (Getz and Shreck, 2006).

Rising commercialism has also created discord among the actors within social movements. Edwards (2013), who has examined changing processes within the Indonesian organic movement, showed that two prominent Indonesian groups engaged in the movement managed to retain the movement's original visions despite an overall shift toward commercialization and institutionalization. One group avoided being co-opted by conventional agriculture because of its close affiliation with a radical transnational peasant movement. However, the other group continued to critique the conventional agricultural system, emphasizing the benefits of organic agriculture as a deliberate strategy to increase sales of its own organic products.

In this chapter we present another case entailing the retention of the original visions of organic and Fair Trade initiatives and avoidance of the typical contradiction that prevails between movements and certification, even in a context of expanding certification-based businesses. This case study centers on organic rice production in Thailand. Unlike the Indonesian case studied by Edwards (2013), Thai rice producer groups, discussed here, neither attempted to garner publicity for their organic products by attacking conventional farming nor maintained strong ties with any international peasant movements. Instead, they were closely aligned with longstanding, homegrown social and environmental justice movements that were also supported by traditional values, most notably Buddhism. Organic agriculture and Fair Trade, both of which are of Western origin, have merged within these existing movements without creating apparent contradictions between foreign and local ideologies, or between organic and Fair Trade initiatives. This raises the question of how Western-derived organic agriculture and Fair Trade concepts have been harmonized with existing Thai social movements. This chapter aims to explore this question.

The following section reviews the development of organic agriculture and examines how the Fair Trade initiative was merged with the organic movement in Thailand. The section also elucidates Buddhist-led social and environmental movements that are unique to Thailand as the background tapestry into which foreign-derived notions and practices of organic farming and Fair Trade have been woven. Subsequent sections sketch outlines of selected cases, namely, three farmer groups located in two provinces of Northeast Thailand, focusing on each group's process of organizational development. Based on a comparison of the three farmers' groups, the remaining sections describe how some of the groups have successfully retained the original visions of social movements. Except where otherwise noted, this chapter describes the situation up to 2013 based on data collected by Tsuruta, the second author, in northeastern Thailand, mainly in the provinces of Surin and Yasothon, between 2012 and 2016.

5.2 The experience of Thailand: An overview

5.2.1 Development of organic agriculture

Until recently, the top-down approach adopted by centralized state agencies has played a limited role in promoting sustainable agriculture (Amekawa, 2010; Kasem and Thapa, 2012), with private initiatives having been instrumental in promoting Thai organic agriculture. During the initial stage of the movement a variety of locally based activists, including farmers, Buddhist monks, and NGOs, constituted the main driving force behind the sustainable farming agenda. They were connected through a nationwide network known as the Alternative Agriculture Network (AAN), which has been promoting alternative forms of farming from the 1980s and onward to counteract the negative social and environmental impacts of rapid agricultural modernization (Vitoon, 2012; Withuun, 1996). In the mid-1990s there were over 50 NGOs working to promote sustainable agriculture across Thailand (Nitasmai, 1996).

By the 1990s the term "alternative agriculture" had come to encompass several revolutionary sustainable farming methods in contrast to the current technical (and politically neutral) understanding of standardized organic agriculture based on certification systems. There were two central underlying concepts during this period. Initially (in the mid-1980s), "integrated farming" was promoted by AAN activists as a concept that directly opposed the specialized monoculture of conventional agriculture. Integrated farming denotes a method entailing the combination of several different agro-ecosystems (such as paddy fields, orchards, and ponds) with diverse crops to conserve ecological integrity while achieving food security. Another concept promoted by the AAN was "natural farming," advocated by Masanobu Fukuoka, a Japanese practitioner-cum-philosopher. This entailed the application of unusual techniques, including zero-tillage and non-weeding (Withuun, 1996). Both concepts strongly emphasize securing farmers' self-sufficiency and self-reliance in conjunction with an ecological balance. At the same time the concept of "alternative agriculture" also implied the ambitious goal of building a more equitable socioeconomic system by protecting small farmers from capitalist exploitation (Withuun, 1996), thereby also evoking Fair Trade principles.

The AAN's grassroots campaign for promoting alternative agriculture was closely linked with other strands of social activism: in particular nationwide environmental and peasant rights movements, which were most active in the northeast, considered Thailand's poorest region (Missingham, 2003; Somchai, 2006). The AAN's lobbying played an important role in achieving the inclusion, for the first time, of the clauses on sustainable agriculture in the seventh National Economic and Social Development Plan in 1997. The concept of sustainable agriculture formulated in the Plan was, in retrospect, radical in that it included subsistence-oriented (rather than commercial) farming styles such as natural farming and integrated farming, reflecting the influence of AAN activism (Vitoon, 2001; Withuun, 1996). However, the potentially radical term "alternative agriculture" was not adopted in this and subsequent national development plans, which demonstrated a preference for more politically neutral and acceptable terms such as sustainable agriculture and organic agriculture.

The sustainable agriculture agenda, which was initially proposed as an alternative to conventional ways of farming and trading, was gradually subsumed into the commercial mainstream. Especially after the introduction of various certification systems in the 1990s, the term "organic" came to be used more frequently as a concept that was closely linked to certification and standardization (Vandergeest, 2009). From the early 2000s and onward, certification-based marketing of organic products became a common practice, considerably boosting certified organic production.[1] Natedao (2011a) aptly summarized this change as marking a transition from alternative agriculture to market-oriented sustainable (or organic) agriculture visions. Alternative agriculture, focusing on promoting the self-sufficiency of small farmers in combination with ecological integrity, has now been replaced by commercial organic agriculture based on standards that are determined by international regulations and labeling systems.

Some AAN members have played an important part in the institutionalization and commercialization of organic agriculture. Specifically, they have played a key role in the establishment, in 1995, of a non-governmental certifying body known as Alternative Agriculture Certification Thailand. Although this institution initially aimed to certify products for domestic markets, the founders soon switched their focus to international organic standards. The organization was consequently renamed Organic Agriculture Certification Thailand (ACT) in 1998, thus discarding the term "alternative" in the original name (Natedao, 2011a; Vandergeest, 2009). ACT was eventually accredited in 2001 by the International Federation of Organic Agriculture Movements (IFOAM), an international organization responsible for setting organic standards. A parallel initiative launched by some AAN activists in 1994 was the establishment of a pioneering organic and Fair Trade organization called the Green Net Cooperative in Thailand's capital city, Bangkok (Vandergeest, 2009; Vitoon, 2001).[2] The Green Net Cooperative has played a pivotal role in promoting export-oriented certified organic rice production in the northeast (Aarat, 2013).

The establishment of a commercial organic certification scheme and associated export businesses coincided with the increasing role of the government in promoting organic agriculture. With its political leverage gained through a high-profile mass protest that occurred in 1997, the AAN secured a promise from the government to support sustainable agriculture, which was later realized as a three-year pilot project from 2001 to 2003 to promote sustainable farming (Vitoon, 2001). This project marked an important turning point toward the increased role of the government in promoting organics. In addition, from 1999 and onward, a social investment fund (SIF) obtained from the World Bank began to be dispensed to villages in the northeast to support grassroots organizations, including some farmer groups cultivating organic rice (Shigetomi, 2010). These two projects helped some farmers' groups to obtain a hefty initial sum of capital for establishing industrial rice-milling facilities, which they could not otherwise have afforded.

5.2.2 *Fair Trade initiative*

Thailand's Fair Trade initiative dates back to the 1970s when some Christian organizations began to trade handicrafts made by ethnic minorities living in the mountainous regions of northern Thailand (Le Minoux, 2012). Subsequently, in the 1990s the initiative acquired momentum in response to rising demands for Fair Trade products, especially agricultural products like coffee and jasmine rice within the global North (Pendergrast, 2015).

In northeastern Thailand, the Fair Trade initiative has been closely linked to the organic agriculture movement, making it difficult to separate one from the other. One of the first agricultural products sold in the Fair Trade markets was rice, which was exported from Surin Province in this region in 1991. This export venture eventually led to the establishment of the Green Net Cooperative (which was also registered as a rice exporter) by AAN members. The export of organic rice to ethical traders in Europe gradually increased, and the Green Net Cooperative received Fairtrade certification in 2002. Several farmers' groups were organized as primary producers under the Cooperative (Green Net, 2016). Following this initiative, obtaining both organic and Fairtrade certification became an essential prerequisite for establishing a successful organic jasmine rice export business. This entwining of the organic and Fair Trade initiatives can also be observed in the ideologies articulated by local farmers and activists that are heavily influenced by the prevailing Buddhist tradition.

5.2.3 *Influence of Buddhism*

Because Thailand is a predominantly Buddhist country, it is inevitable that various rationales concerning social ethics will invoke Buddhist ideas and values. In the aftermath of the brutal suppression of political activism in the mid-1970s, Thailand witnessed a rise in non-violent civic movements advocating social (and later environmental) justice in line with "socially engaged Buddhism" (Gohlert, 1991; Nishikawa and Noda, 2001).[3] Buddhadasa, a heretical monk and scholar, was one of the most influential Buddhist thinkers and provided an ideological framework for reformist intellectuals. He conducted a radical and fundamental examination of society based on *dhamma* (also spelled *dharma* in English literature), which is Buddhist truth or law. The core of Buddhadasa's ideas on social reform has been summarized by Taylor (1997, p. 40) as being "the need for living a simple and satisfying life in harmony with nature as a counter-force to the inequities and injustices of modernization."

Followers of Buddhadasa interpreted his teachings, especially those on *dhamma*, as supporting contemporary principles of social and environmental justice, which were later manifested in organic and Fair Trade initiatives. *Dhamma* is an ambiguous term translated variously as truth, virtue, or cosmic (natural) law. Buddhadasa argued that the meaning of *dhamma* included nature itself, the laws of nature, and duty in accordance with the laws of nature (Ito, 2012, p. 193). Accordingly, a society based on *dhamma* (interpreted as natural law) should be an egalitarian society consonant with *dhamma* (interpreted as virtue), because economic exploitation is absent in a pure, original, natural world wherein all living things are interdependent (Buddhadasa, 1986). Thus, both ecology and social justice principles, which are often seen as being distinct in Western thought, can be derived from the single notion of *dhamma*. Buddhadasa is likely to have influenced the popularization of the term *dhamma*-ness as a key concept denoting social and political justice within activist circles in the 1970s and 1980s (Ito, 2012). The indirect influence of Buddhadasa is evident in the term *kaan khaa thii pen tham* (trade based on *dhamma*-ness), which is the Thai equivalent of Fair Trade.

Organic agriculture is also considered to be closely connected with *dhamma*. An organic farmer the author met in Surin Province noted:

> You cannot practice organic farming if you don't have *dhamma*. If you have *dhamma*, you can reflect it through your conduct and turn your thought[s] to other living things. You understand that all creatures depend on each other. Killing fish by chemicals is a sin. Living creatures in paddy fields such as shrimps and mice help us in breaking and enriching the soil.
>
> (November 4, 2014)

Along with the *dhamma* concept, the Five Precepts, referring to a familiar religious discipline followed by lay Buddhists, are often invoked in connection with the practice of organic farming.[4] As indicated in the farmer's statement here, the First Precept, that is, not killing living things, is often cited to justify organic farming practices (Kaufman, 2012; Kaufman and Mock, 2014).

The inseparability of the organic and Fair Trade movements has also been evident in the activities of rural activist monks, dubbed "development monks," who have initiated and coordinated various community development programs in northeastern villages since the 1970s. Working closely with the AAN and other NGOs, these activist monks have passionately sought to improve the livelihoods of villagers in this region, who have endured chronic indebtedness and environmental degradation. Both problems have been partially caused by the extensive application of chemical fertilizers and pesticides. Pursuing the ideal of building a society based on *dhamma*, development monks have tried to find ways to lift farmers out of poverty and to persuade them to share community resources including rice and cash, as well as the wealth of their local knowledge. Various measures implemented to foster economic self-reliance included the establishment of rice banks[5] and community savings banks, as well as the promotion of folk knowledge, particularly that on sustainable farming and medicinal herbs (Pinit, 2012; Seri, 1988).

Although these monks and their associates did not use the term Fair Trade, their activities were in alignment with Fair Trade principles and practices. In particular, one of their main aims was to overcome inequity in the transactions of rice, which is a staple food crop as well as an important cash crop in Thailand. In the 1970s and 1980s indebted farmers who were hard up for cash were forced to sell their paddy immediately after harvesting it when the price was at rock bottom, and subsequently had to purchase or borrow rice for consumption at a higher price (Phittaya, 1993). Local rice markets were mainly controlled by urban-based owners of commercial rice mills who were notorious for cheating farmers and beating down the price of their paddy. Therefore, rice collected and stored in community rice banks and farmers' own rice mills became important assets for regaining control over the farmers' primary food crop, as well as for increasing their bargaining power in rice sales. Following the introduction of the concept of integrated farming in the mid-1980s, development monks helped to promote integrated farming methods based on *dhamma*. As discussed later in this chapter, with the facilitation from AAN-affiliated activists such rice banks evolved into businesses entailing rice collection for domestic and international markets.

In sum, the Thai alternative agriculture movement, which appears to be based on foreign ideas such as organic and natural farming methods, is also grounded in indigenous Buddhist values in terms of both its ideology and its practice. This is aptly conveyed in the term "Buddhist agriculture," advocated by the influential social critic and activist Prawase Wasi, who developed an interpretation of alternative agriculture and associated social reforms in Buddhist terms. He envisaged a harmonious and balanced rural life based on Buddhist traditions (especially the spirit of *dhamma*) and integrated farming to enable farmers to become self-reliant and disengage themselves from the vices of consumerism (Prawase, 1988). He provided a lot of practical and moral support for development monks and the AAN (Parnwell, 2005; Withuun, 1996).

5.2.4 Organic and Fairtrade rice as an emerging export sector

This case study on Thailand focuses specifically on rice, which is one of many organic and Fair Trade products. Rice is not only a staple food for Thai people, but it has also been one of the country's flagship export items. Although the importance of rice (and of the agricultural sector as a whole) in Thailand's exports has considerably declined, in combination with processed rice products it remains the second most widely exported item after natural rubber in the agricultural sector. In 2013 rice accounted for approximately 12 percent of the total value of agricultural exports (Office of Agricultural Economics [OAE], 2014). Commencing in the 1980s, the production of improved varieties of aromatic rice, known as upmarket jasmine rice, for both domestic and international markets has been steadily rising, particularly in the northeast. Within a short period the country's most impoverished region has emerged as one of its key centers for jasmine rice production. Jasmine rice accounted for 44 percent of Thailand's total export value of rice (excluding processed rice products) in 2013 (OAE, 2014).

As Figure 5.1 shows, whereas other organic crops have rapidly expanded in Thailand, organic rice (mostly of the jasmine varieties) far exceeds them in terms of its importance among Thailand's organic agricultural products. Jasmine rice has also been the most important Fairtrade item exported from Thailand.[6] The first organic rice project in Thailand was launched in 1989 by a private company,

(Unit: 1,000 hectares)

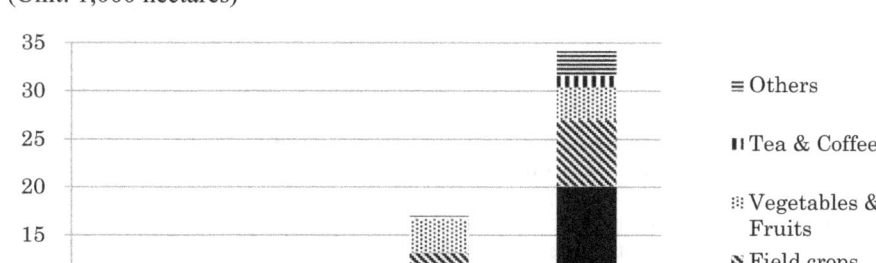

Figure 5.1 Organic-certified areas by crop in Thailand

Source: Vitoon (2015, p. 2)

Note: The data for rice in 1998 and 2003 include data on field crops (e.g., sugarcane and cassava).

which initiated contract farming with some upland farmers in northern Thailand in response to a request made by an Italian trader (Vandergeest, 2009). Attracted by the commercial viability of organic jasmine rice, a number of farmers' groups and cooperatives subsequently began to produce it for export. Currently, most of Thailand's organic rice is produced in the northeast in particular, as well as in the north. As Table 5.1 shows, the northeastern provinces of Yasothon, Ubon Ratchathani, and Surin are the main organic rice-producing areas. The same table also indicates that organic rice production in the northeast is characterized by small landholdings as compared to the central region, in which organic farming is practiced on much larger farms and plantations.

The existing literature suggests that organic and Fair Trade initiatives generally bring about positive economic benefits for jasmine rice producers, although resource-poor farmers find it difficult to survive during the conversion period. Much of the literature on organic jasmine rice production in the northeast tends to focus on the economic and technological differences between organic and non-organic farmers in the area. Most of these studies provide evidence of the economic as well as social benefits of organic farming and Fairtrade certification in the northeast because of the premium prices of organic or Fairtrade rice (Becchetti, Conzo, and Gianfreda, 2009; Manas and Prasit, 2007; Nuntana and Winnett, 2002; Sununtar, Leung, and Cai, 2006). Some studies have revealed a variety of challenges faced by converts to organic farming, especially among poorer sectors of village communities. Based on her in-depth anthropological research on an organic farmers' group, Natedao (2011b) argued that while enjoying access to niche markets in the global North, organic rice farmers faced

Table 5.1 Top five provinces engaged in organic cultivation in Thailand, 2013

Province	Region	Major crops	Organically cultivated area (hectares)	Number of households engaged in organic cultivation	Average area per household (hectares)
Nakhon Pathom	Central	Coconuts	5,630	199	28.3
Yasothon	Northeast	Rice	5,071	1,375	3.7
Ubon Ratchathani	Northeast	Rice	4,639	1,475	3.1
Chiang Rai	North	Rice and vegetables	3,769	922	4.1
Surin	Northeast	Rice	3,212	1,206	2.7

Source: Vitoon (2015, p. 3); personal communication with Vitoon (April 21, 2016)

several challenges, including the need for intensive labor and stringent international standards. She further observed that the majority of farmers lacked the technical and financial support required for converting to organic agriculture, which could therefore give rise to a new type of rural social differentiation. Nuntana and Winnett (2002) found that poorer farmers faced difficulties during the period of conversion from non-organic to organic agriculture, during which they were unable to obtain full premium prices despite the risk of crop failure. However, fully organic rice producers in the northeast obtained both economic and social benefits from a Fair Trade project.[7]

5.3 A focus on three groups of rice producers

5.3.1 Two categories of farmer groups

Recently there has been increasing competition among organic jasmine rice producer groups in the northeast. The proliferation of organic rice producer groups largely reflects increasing sales opportunities for organic jasmine rice, not only in Europe and North America but also in parts of Asia such as Hong Kong, Singapore, and mainland China. Organic rice producer groups can be divided into two categories. As Table 5.2 shows, some farmers' groups began producing organic rice even before ACT (the local certifying body) began implementing inspections in the late 1990s (see Subsection 5.2.1). On the other hand, most producer groups began cultivating organic rice in the 2000s when organic certification became quite common as a result of an increasing number of foreign certifying bodies operating in Thailand. The former category, that is, farmer groups that initiated organic production during the pre-ACT period, seem to differ from most of the groups belonging to the post-ACT category in terms of their commitment to long-ranging social objectives.

Table 5.2 Major producers of organic rice in northeastern Thailand, June 2016

Category	Group name	Province	Start of organic rice production (estimate)	Annual organic and pre-organic paddy collected from members (estimate)	Dominant concepts of alternative agriculture
Pre-ACT period	**Rice Fund Surin Organic Agriculture Cooperative**[a]	Surin	1990	741 tons (2011–2012)	Integrated farming and natural farming
	Nature Care Society (NCS)[b]	Yasothon	1990	663 tons (2011–2012)	
	Bak Ruea Farmers Group[b]	Yasothon	1996	480 tons (2011–2012)	
	Nong Yo Natural Organic Agriculture Cooperative[c]	Yasothon	1990s	100 tons (2011–2012)	

The first local organic certifying body in Thailand, Organic Agriculture Certification Thailand (ACT) started inspection and certification of organic agricultural products during 1997–1999.

Category	Group name	Province	Start of organic rice production (estimate)	Annual organic and pre-organic paddy collected from members (estimate)	Dominant concepts of alternative agriculture
Post-ACT period	**Phathana Community Enterprise Network (pseudonym)**[a]	Yasothon	2000	2,100 tons (2011–2012)	Standard-based organic farming
	Organic Agriculture Cooperative Surin[a]	Surin	2001	1,500 tons (2013–2014)	
	Organic Jasmine Rice Producer Group (Progressive Farmers Association)[a]	Ubon Ratchathani	2001	3,200 tons (2011–2012)	
	Loeng Nok Tha – Thai Caroen Organic Agriculture Cooperative[b]	Yasothon	2004	200 tons (2011–2012)	
	Than Lux Cooperative (Thai Smart Life)	Yasothon	2004	100 tons (2010–2011)	
	Moral Rice Project[c]	Yasothon	2005	250 tons (2011–2012)	
	Prasaat Agricultural Cooperative	Surin	2006	59 tons (2011–2012)	
	Amnat Charoen Organic Agriculture Community Enterprise Network[a]	Amnat Charoen	2008	600 tons (2014–2015)	
	Ban Um-sang Rice Community Enterprise[a]	Sisaket	2009	3,000 tons (2013–2014)	
	Nong Yang Organic Rice Growers Group[c]	Yasothon	2009	200 tons (2014–2015)	
	FTS Organic and Fairtrade Producers[a]	Amnat Charoen	2010	N/A	
	Sachatham Rice Project	Amnat Charoen	2011	1,100 tons (2014–2015)	

Source: Field surveys

Notes: Existing organic farmers' groups in the northeast are not exhaustively covered in the above table.
[a] Direct Fairtrade certification (as of 2013).
[b] Fairtrade certification under the name of the Green Net Cooperative (as of 2013).
[c] These groups originate from NCS.

The remaining sections of this chapter highlight the differences between these two categories of producer groups in terms of their economic and social characteristics through an examination of three farmer groups: two long-established farmer groups, namely, the Rice Fund Surin Organic Agriculture Cooperative (hereafter Rice Fund Surin) and the Nature Care Society (hereafter NCS) belonging to the pre-ACT category; and one group, Phathana Community Enterprise Network (hereafter Phathana Enterprise) belonging to the post-ACT category.

Data on these three groups were obtained in the provinces of Surin and Yasothon through interviews conducted with about 50 key informants, including members (farmers and managers of Rice Fund Surin, NCS, and Phathana Enterprise as well as their board members), one development monk, and provincial officials in charge of agriculture and commerce. These interviews were complemented by visits to members' farms, participation in seminars and meetings organized by the groups and related organizations, and analysis of annual reports and other documents published by the three groups, the AAN, and an NGO supporting Rice Fund Surin.

5.3.2 History of the farmer groups

5.3.2.1 Rice Fund Surin Organic Agriculture Cooperative (Rice Fund Surin)

The inception of Rice Fund Surin dates back to the 1970s when Luangpho Naan, a Buddhist monk, initiated various development projects aimed at improving the livelihoods of the poor within a village located in proximity to the city of Surin, the capital of Surin Province (Figure 5.2). The monk, who later became known nationwide as a prominent development monk, established a rice bank in 1981 to secure villagers' control over their staple food. Besides deploying cash and rice donated to his temple as initial capital for the bank, he patiently persuaded villagers, especially wealthier ones who had doubts about the rice bank project at the outset, to deposit their paddy in the bank as a *dhamma*-based act (Izumi, 1995). The rice bank project achieved notable success in alleviating rice shortages and indebtedness among poorer households. With the inclusion of neighboring villages that immediately followed suit, an inter-village network of several rice banks cooperating with each other was formed. Luangpho Naan also launched a cooperative shop in 1982 and a savings union in 1985 in his village (Izumi, 1995; Phittaya, 1993). The funds within the savings union were loaned to some union members who needed to pay off their debts and begin integrated farming with the aim of attaining food self-sufficiency, while other members took loans to buy chemical fertilizers (Nozaki, 1995; Seri, 1988).

With the help of external activists, the monk and local farmers launched a network, called Surin Farmers Support (SFS), to empower the rural poor across Surin Province. SFS, which was connected to the nationwide network of the

Figure 5.2 The late Luangpho Naan (right), photographed by the author in
    November 2012

AAN, promoted integrated and natural farming as one of its important rural
development strategies. During the late 1980s, several natural farming groups
formed in different locations within Surin Province attempted to initiate a joint
rice marketing enterprise with the support of SFS. Rice from Surin Province was
first sold to ethical trade shops in Bangkok through the AAN network. In 1991,

for the first time, rice was exported to OS3, an NGO based in Switzerland and a predecessor of Claro Fair Trade. Working with SFS, Luangpho Naan played a central role in accumulating rice obtained from existing rice banks within several villages for export to Switzerland (Nozaki, 1995). Over time OS3 requested that farmers in Surin Province produce completely organic rice by the mid-1990s. During this process the original natural farming method, which failed to bring the expected yields to farmers, was gradually replaced by organic farming entailing the use of animal manure and nitrogen-fixing crops.

SFS gradually shifted its priorities from promoting integrated or natural farming to developing a rice business. Rice Fund Surin took shape in 1992 as the marketing section of one of the natural farming groups. Initially Rice Fund Surin had neither a permanent office nor a rice mill. Luangpho Naan offered up the use of his temple as a packing facility and installed a secondhand rice mill near the temple (Aarat, 2013). Rice from Rice Fund Surin was exported through the Green Net Cooperative (see Subsection 5.2.1), and ACT initiated inspections for organic certification in the late 1990s. In 2003 Rice Fund Surin was officially registered as an agriculture cooperative. Because orders received from the Green Net Cooperative were unable to keep pace with the increasing supplies obtained from its members, Rice Fund Surin decided to develop its own export business network, eventually registering itself as an independent exporter and directly receiving Fairtrade certification in 2005 (Aarat, 2013). The cooperative acquired additional organic certification for the European Union (EU) and the National Organic Program of the United States (NOP).[8]

As a final step, SFS, which had hitherto not been a permanent organization, was renamed the Community of Agro-Ecology Foundation (CAEF) and officially registered as an NGO in 2009. In other words, SFS gave birth to "twins," each with its own specialty: Rice Fund Surin catered to the business sector, and CAEF promoted sustainable agriculture and empowered rural communities as a non-profit organization. These twin organizations, located in neighboring building sites, have been working in close collaboration with each other.

5.3.2.2 Nature Care Society (NCS)

Similar to the case of Rice Fund Surin, NCS's origins lie in several intertwined community development activities implemented in several villages in Yasothon Province in the 1980s. These activities entailed collaboration between villagers, external activists, and another local development monk named Phra Khru Suphacharawat. In 1983 Phra Khru Suphacharawat, in conjunction with some villagers, organized an association of villagers to promote traditional herbs for achieving self-reliance in medical care, with assistance provided by a Bangkok-based NGO (Nuntiya, 1998). The association not only established an herbal medicine center at the monk's temple, but also launched projects for community forest rehabilitation and the promotion of alternative agriculture. The monk actively promoted integrated and natural farming methods among local farmers, based on his staunch faith that these farming methods accorded with the teachings of *dhamma* (Parnwell and Seeger, 2008). He first introduced these farming methods as a pilot project implemented in farmland owned by one of his relatives and demonstrated the usefulness of the methods to neighboring farmers who had doubts about this experiment (Nishikawa and Noda, 2001). In 1990 Masanobu Fukuoka, the guru of natural farming, was invited by organizers of the herbal project to give a lecture attended by local farmers. Among the attendees, those who were deeply inspired by Fukuoka's lecture established a farmers' group, subsequently named the Nature Care Society (NCS), with the aim of promoting natural and integrated farming, as well as the economic self-reliance of local farmers.

Members of NCS sought to construct their own rice mill to increase their bargaining power during price negotiations, as well as to promote sales of chemical-free rice. They finally succeeded in constructing a mill in 1991 as a result of a unique instance of cooperation between farmers and urban consumers. Apart from a small contribution made by each villager, some activists and consumers in Bangkok established an association of consumers to provide a financial support base for the construction of the rice mill.[9]

By the mid-1990s NCS had joined the rice-exporting business network of Rice Fund Surin and the Green Net Cooperative. NCS has continued to retain close business relationships particularly with the Green Net Cooperative. In the late 1990s ACT, or organic certification for export, was introduced, and a second rice mill of greater production capacity was established with financial support obtained from SIF in 1999 (see Subsection 5.2.1). Other facilities, including a large storehouse, were added with the receipt of financial aid from the government (Walaiporn, Areerat, and Manitchara, 2007). To utilize a relatively large milling capacity, NCS accepted organic as well as non-organic rice from its members (Nuntiya, 1998).

5.3.2.3 Phathana Community Enterprise Network (Phathana Enterprise)

While the decade following 2000 witnessed a proliferation of commercial organic farmer groups in northeastern Thailand (see Table 5.2), unlike Rice Fund Surin and NCS, many of these did not originate in the rural social activism that evolved in the 1980s. One of these groups, Phathana Enterprise, located in Yasothon Province, provides an example of business-oriented producer groups that operate for commercial success rather than to attain the original visions of alternative agriculture.

Phathana Enterprise originated in a community development program that was supported by the Thai government and an NGO in 1999. The program aimed to encourage villagers to design a development plan by themselves for the Phathana sub-district (pseudonym), comprising 12 villages.[10] As a result, villagers identified the rising costs of agricultural chemicals as one of the overriding problems in the area and subsequently decided to promote sustainable agriculture throughout the entire sub-district, which led to the formation of an unregistered organization known as the Phathana Group for Sustainable Agriculture.

From the outset the Phathana Group's activities seem to have been devoted exclusively to the business of exporting certified organic jasmine rice. In 2002 the group started to pursue organic certification and to simultaneously attempt to produce compressed organic fertilizer. Even during the conversion period they exported their rice to some Asian countries. In 2004 the group constructed a rice mill with financial support from the government. By 2006 the group began to export fully organic EU-certified rice verified by a German certification body. The group also obtained Fairtrade certification in 2004.

Also in 2002, just two years after its inception, the Phathana Group comprised 250 members, mostly from the same sub-district. Attracted by the premium price of organic jasmine rice, new members joined the group and began producing this item. Ten years later, the number of members had expanded to 564. Thus, within a short period the group became the unit producing the largest quantity of organic rice in Yasothon Province (see Table 5.2). Consequently, approximately half of the paddy fields in the Phathana sub-district were converted for organic cultivation by 2012. By 2009 the group was officially registered as a community enterprise network, which made it possible to secure a loan from the Bank for Agriculture and Agricultural Cooperative, a government bank, to construct a new rice mill in 2010. In 2011, to prove the high quality of their jasmine rice, Phathana Enterprise also obtained a geographical indication for their product, complying with EU regulations.[11]

5.4 A comparison of the three farmers' groups

5.4.1 Organizational processes

The historical backgrounds of the two pre-ACT farmer groups (Rice Fund Surin and NCS) and of the single post-ACT group (Phathana Enterprise) differ. Rice Fund Surin and NCS owe their origins to various community development projects for fostering self-reliance, based on the interactions of heterogeneous actors including villagers, development monks, and urban activists affiliated with the AAN. Such grassroots activities, in addition to a process of trial and error that unfolded in the search for alternative agriculture, gradually converged in the form of standardized organic farming as a response to international regulations. By contrast, from its inception Phathana Enterprise embarked on certified organic agriculture with the singular aim of expanding its export business. The focus on business may have been a reason why this group accumulated a sizable membership so rapidly. Another reason behind the rapid expansion of Phathana Enterprise seems to have been its geographical location almost within one sub-district. It was thus comparatively easier for Phathana Enterprise to recruit members. Unlike the membership of Phathana Enterprise, that of Rice Fund Surin and NCS comprised farmers from several different districts within each province.

These different historical backgrounds are also reflected in the present organizational structures of the three groups. In contrast to Phathana Enterprise, which was formed through a top-down process mainly for business purposes, Rice Fund Surin was established as an amalgamation of several different sub-groups of local farmers connected within the wider network of SFS (see Subsection 5.3.2.1). Each component group within Rice Fund Surin conducts its own diverse activities that are appropriate for its local context. Rice Fund Surin can be regarded as a federation comprising a number of independent local rice grower groups. It is also closely associated with SFS's successor, CAEF, which is a non-profit organization working to promote sustainable rural development.

One characteristic that is unique to NCS is that its membership includes both organic and non-organic producers, and even non-producers such as consumers who live outside of Yasothon Province in other provinces and in Bangkok (NCS, 2012). NCS accepts not just organic paddy for export, but also non-organic paddy for domestic markets, including a small amount of glutinous rice, which is the staple food of the local population, to fulfill the diverse needs of local farmers and consumers. The inclusion of consumers within NCS's membership reflects its unique history of collaborative engagements between producers and consumers (see Subsection 5.3.2.2). The majority of its members who live in Bangkok are people who supported the construction of NCS's first rice mill, and they have subsequently retained their shares in the mill over the last two decades. Some of these members still purchase organic rice from NCS. Furthermore, unlike the other two groups, NCS has maintained close business relationships with the Green Net Cooperative and ACT, both of which are offshoots of AAN activism.

Lastly, the three groups also differ in terms of their legal statuses, as they are registered differently as an agricultural cooperative, a farmers' association, and a community enterprise network (see Table 5.3). Although they all enjoy the privilege of a tax reduction and can receive financial support from governmental institutions, only Rice Fund Surin, a cooperative, can register itself as an independent exporter.

5.4.2 Sales of certified organic rice

Notwithstanding the different historical backgrounds of the three groups, they do share basic common characteristics: (a) they are all organizations for smallholder farmers, (b) they are engaged in the processing and selling of rice, and (c) they are in compliance with international organic and Fairtrade standards (see Table 5.3). Organic–Fairtrade double certification is now an integral part of their export businesses in response to the requests of a number of their foreign (mostly Western) customers. For instance, it is estimated that at least 50 percent of rice purchased by Rice Fund Surin and Phathana Enterprise was exported with both kinds of certification during the 2011–2012 season.[12] All of the groups acknowledged that the double certification greatly helped them to build up a thriving rice business, significantly enabling easier access to the niche market in the global North through certification as well as the advances and Fairtrade premiums that

Table 5.3 Summary of the three producer groups in northeastern Thailand

	Rice Fund Surin with its sister NGO, CAEF	NCS	Phathana Enterprise
Rice mill business			
Current legal status	Agricultural cooperative (since 2003)	Farmers' association (since 1976)[d]	Community enterprise network (since 2009)
No. of members	267 (organic, pre-organic, and preliminary stage[a])	221 (organic and pre-organic); 685 (non-organic); and 110 (consumers)	564 (organic and pre-organic) in 2012–2013
Average area of organic paddy fields per member (estimate)	3.5 hectares (organic, pre-organic and preliminary stage[a])	2.4 hectares (organic and non-organic)	3.1 hectares (2012–2013) (organic and pre-organic)
Amount of paddy purchased from members (estimate)	741 tons (organic, pre-organic, and preliminary-stage jasmine rice[a])	663 tons (organic and pre-organic jasmine rice); 1,000 tons (non-organic and non-glutinous); and 20 tons (glutinous)	2,100 tons (organic and pre-organic jasmine rice)
International organic certification	IFOAM (since 2001); EU and NOP (since the mid-2000s); and Bio Suisse (since 2011)	IFOAM (since 2001); EU (since 2011); and Canada Organic Regime (since 2012)	EU (since the mid-2000s); NOP (since 2011)
Organic certifying agency	ACT (since the late 1990s); and BCS Öko-Garantie (since the mid-2000s)	ACT (since the late 1990s)	BCS Öko-Garantie (since the mid-2000s)
Fairtrade certification	Certified under the name of Green Net Cooperative since 2002 and directly certified since 2005	Since 2002 (certified under the name of Green Net Cooperative)	Directly certified since 2004
Exporting agency	Green Net Cooperative until 2005; and Rice Fund Surin (registered as an exporter) since 2005	Green Net Cooperative	Private exporting companies (excluding Green Net Cooperative)

	70% to 80%	40 %	85% to 95%
Proportion of rice for export[b]	70% to 80%	40 %	85% to 95%
Major destinations of export	EU countries, USA, Australia, Switzerland, and Singapore	EU countries	EU countries and USA
Packing facilities	Own	Owned by the Green Net Cooperative	Own
Activities other than rice-based business — Community bank	Available (for members only)	Available (for both members and non-members)	Not available
Welfare fund[c]	Not available	Available	Not available
Sales in local organic markets ("green markets")	Active	Active	Not involved
Domestic consumer network	Available	Available	Not available
Preservation of traditional varieties of rice	Active	Active	Not active

Source: Field surveys, 2012 and 2013

Notes: The data are from the production year 2011–2012, unless otherwise indicated.

[a]During what is recognized as a "preliminary stage," which corresponds to the inception period of organic farming, Rice Fund Surin permits a producer to retain both organic and non-organic fields.

[b]Calculated as the ratio of organic paddy for export to the total quantity of paddy, both organic and non-organic, collected by each group.

[c]Excludes Fairtrade premiums.

[d]When the first rice mill was constructed in 1991, the group decided to use the name of an existing but non-functional farmers association that had been formed for the distribution of chemical fertilizers in the 1970s.

they received. Whereas the leaders of Rice Fund Surin and NCS tend to stress their commitment to the ethical requirements of Fairtrade certification, Phathana Enterprise staff regards Fairtrade certification simply as a convenient scheme for exporting pre-organic rice that cannot be exported to Europe and the United States as organic-certified rice. During the 2011–2012 season roughly 30 percent of paddy purchased by Phathana Enterprise was obtained from fields undergoing a process of conversion, with most being exported as a non-organic Fairtrade product.

All three groups first pursued organic certification and were then certified as Fairtrade producer groups. From the early 2000s and onward, their rice production and marketing operations have always been accompanied by both organic and Fairtrade certification. However, the members of these groups basically consider themselves to be organic rather than Fairtrade producers, partly because of the great deal of work that the organic certification process necessarily entails. Apart from official annual inspections made by the certifying agencies, organic certification requires a laborious operational process of internal control over a whole cropping season, which includes conducting several field surveys by internal inspectors and the maintenance of detailed records and documents at both the levels of the individual farm and of the entire group. In stark contrast, the process for annually renewing Fairtrade certification is almost imperceptible for ordinary members because it takes just a few days without requiring the extensive involvement of the members. The payment of certification fees to the respective certifying agencies is a factor that is common to both organic and Fairtrade certification systems. These payments pose a considerable burden on farmers' groups. In the case of Phathana Enterprise, around 11 percent of the net profit accrued from its rice business was allocated to meet the annual cost of certification in 2012.

Members of each group retain a portion of their products for family consumption, in addition to the seeds used for planting during the following year, with the remaining portion being sold to their associated farmers' groups. However, if the price offered by their farmers' groups is not appealing, they can also sell to other private rice mills dealing with non-organic products. Organic farmers' groups once paid organic premium prices that were 15 percent to 50 percent higher than the market price in the 1990s and early 2000s. However, the rate has decreased considerably and has reached a level as low as approximately 5 percent recently, mainly because in 2011 the new prime minister started to promote a controversial rice-pledging scheme more intensely than before. Under this scheme, the government bought paddy from farmers through private rice mills at a price much higher than the market price, irrespective of organic status. The farmer groups were, therefore, forced to offer their members a price that even exceeded the grossly inflated price offered by the government to secure the necessary amount of organic paddy for export.[13] The rise in the farm-gate price of paddy further strained the budget of the farmers' groups under study, all of which had already become heavily indebted as a result of the loans procured from governmental banks and NGOs.

5.4.3 The use of Fairtrade premiums

The three groups have utilized Fairtrade premiums for their own advantage, and that of their members, in a variety of ways. Table 5.4 shows for what purposes two of these groups, Rice Fund Surin and Phathana Enterprise, used Fairtrade premiums in 2011. Both groups used the bulk of the funds to upgrade their facilities for their rice mill businesses. Although details were not available for the third group, NCS, this group also showed the same tendency. About 50 percent of the Fairtrade premium received by NCS in 2008 was invested in improving its rice mill management (Becchetti et al., 2009, p. 27).

A difference can be observed between the pre-ACT and post-ACT farmers' groups. In the pre-ACT Rice Fund Surin group, one quarter of the Fairtrade premium was used to support the activities of local sub-groups (Table 5.4). The apparent reason for this is that formerly Fairtrade premiums paid to Rice Fund Surin were directly allocated to each local unit of farmers within the group in accordance with the unit's production amount. Although Fairtrade premiums are currently managed centrally to prioritize the improvement of business facilities, each sub-group of farmers proposes its own projects and obtains grants for its local needs. Similarly, within NCS 50 percent of the Fairtrade premium was divided equally among extension services for member-farmers and a fund called the Organic Fair Trade Fund (Becchetti et al., 2009). The former included the promotion of organic farming and training members for organic inspection. The latter, combined with other financial sources, was disbursed as loans to members' organic farming. As of 2012 loans of up to 10,000 baht for an individual, or 20,000 baht for a group, were disbursed in accordance with a plan proposed by

Table 5.4 Disbursement and usage of Fairtrade premiums (two groups in northeastern Thailand), 2011

Purpose	Rice Fund Surin	Phathana Enterprise
1. Improvement of rice mill and office facilities	72.0 %	46.9 %
2. Excursions, workshops, and meetings	–	17.0 %
3. Staff expenses	–	12.5 %
4. Improvement of farming system	–	15.6 %
5. Support for projects of local sub-groups	25.0 %*	–
6. Philanthropic activities	–	4.0 %
7. Eco-friendly activities	–	4.0 %
8. Support for disadvantaged members (flood victims and sick members)	3.0 %	–
Total amount (Thai baht)	313,091	2,151,853

Source: Field survey, 2012; Rice Fund Surin (2012, p. 22)

Note: *As shown in Table 5.5, the support for local sub-groups' projects includes construction of a facility for organic fertilizer production (sub-group A); improvement of a community rice mill (sub-group B); support for organic pig-raising (sub-group C); and financial support for organic inspection (sub-group X).

each member or sub-group (e.g., compost-making, cattle-raising, and organic vegetable cultivation).[14] Fairtrade premiums were used not only to enable members to implement their plans, but also to help members who were victims of diseases or natural calamities such as floods and droughts.

By contrast, the post-ACT Phathana Enterprise allocated a large share (29.5 percent) of the premium for business administration, as shown in rows (2) and (3), see Table 5.4. However, only 15.6 percent of the premium was used for the improvement of members' farming systems, with only 8 percent allocated for welfare activities at the community level, such as donations made to schools and temples, sporting activities, and planting trees, as shown in rows (6) and (7), see Table 5.4.

5.4.4 The fundamental difference between the groups

The most significant difference between the pre-ACT groups, Rice Fund Surin and NCS, and the post-ACT Phathana Enterprise is that a variety of social functions exist separately from organic rice production and sales in the case of the former. The rice mill business is only one component of the groups' multiple activities. As Table 5.3 indicates, whereas Phathana Enterprise's activities primarily focus on the export of organic rice, the multifaceted activities of Rice Fund Surin (along with CAEF) and NCS extend beyond the boundaries of the rice business. In Phathana Enterprise, neither a savings association nor a welfare fund has been set up. Further, the group is not involved in other activities such as traditional seed preservation and local organic markets that are frequently observed in relation to the first two groups.

The business-oriented nature of Phathana Enterprise is also evident in its recent attempt to expand its export business to include products other than rice. In 2010 Phathana Enterprise commenced production of organic and Fairtrade-certified soybeans for processing into soy sauce for export by a Thai food conglomerate. In 2015 Phathana Enterprise, along with other organic rice producer groups in the northeast, took a further step toward establishing a joint venture rice-exporting company to gain direct access to foreign importers. Thus, pre-ACT farmers' groups derived from social movements, such as Rice Fund Surin and NCS, are now facing growing competition from these newer and larger organic producer groups within the post-ACT category (see Table 5.2).

5.5 Growing divergence within the groups

The rivalry and discord between business-oriented groups and movement-derived groups can also be observed within individual farmers' groups. There has been evidence of some contradictory practices, even within Rice Fund Surin, between the original members and business-minded new entrants. As of 2012 there were 13 sub-groups of Rice Fund Surin within different localities, each with a distinctive historical background.

Sub-group X joined Rice Fund Surin relatively recently in the late 2000s and rapidly rose to become one of the largest rice producer groups within the cooperative (see Table 5.5). Unlike most of the other sub-groups, sub-group X officially registered itself as a community enterprise in 2007 and has its own large-scale commercial rice mill, organic certificates, and brand name. This sub-group is semi-independent in terms of its rice business, selling only a part of its produce to Rice Fund Surin. For sub-group X, its membership of Rice Fund Surin may be a mere stepping-stone toward the establishment of an independent business network. During the 2012–2013 season, this sub-group sold only 50 percent of the approximately 500 tons of organic jasmine paddies that they had produced to Rice Fund Surin. They sold another 30 percent as a packed rice to other domestic sales channels.[15] Sub-group X was even seeking an opportunity to export their products overseas, bypassing Rice Fund Surin. Sub-group Y (shown in Table 5.5), also under Rice Fund Surin, is registered as a community enterprise and has been trying to sell its own milled rice products, albeit on a much smaller scale than that of sub-group X.

In response to the emergence of sub-groups such as X and Y that are more business oriented, the board members of Rice Fund Surin have recently urged

Table 5.5 Profiles of major sub-groups under Rice Fund Surin, Thailand

Sub-group	Year of formation	Original purpose	Current activities
A	1991	Saving and natural farming	Conservation of indigenous rice varieties; community seed bank; organic fertilizer production; provision of loans to members; lease of green manure seeds to members; and participation in green markets
B	mid-1980s	Natural farming	Operation of a community rice mill; collective production of jasmine rice seeds; production of organic shallot and garlic; conservation of indigenous rice varieties; restoration of community forests; collaboration with a Bangkok-based NGO working on environmental and educational issues; production of organic fertilizers; and provision of loans to members
C	late 1980s	Natural farming	Provision of loans to members; organic pig-raising; and participation in green markets

(*Continued*)

Table 5.5 (Continued)

Sub-group	Year of formation	Original purpose	Current activities
D	1997	Community forest conservation	Promotion of organic pig production and other income-generating activities for villagers with fewer or no paddy fields; participation in green markets; and promotion of community organic markets at the village level
X	early 2000s	Promotion of organic rice production	Marketing of organic rice with the group's own commercial rice mill; certification and brand name; and participation in green markets
Y	2006	Promotion of rice marketing	Marketing of organic rice and processed rice products under the group's own brand name; operation of a retail shop; production of organic fertilizers; and participation in green markets and other farmers' markets

Source: Field surveys, 2012 and 2013

members to fulfill the original contract and deliver at least 70 percent of their produce to the cooperative, limiting the amount to be sold through other market channels outside of the cooperative. It has become more difficult for wealthier farmers to join the cooperative and avail themselves of its advantages. This has helped to preserve the cooperative's original objective, that is, to help small farmers.[16]

5.6 Efforts to strike a balance between business and ethical initiatives

Focusing on the two pre-ACT farmers' groups, namely Rice Fund Surin and NCS, this section further illustrates how, despite the expansion of the rice export business, some sub-groups of farmers have continued to maintain the original alternative agriculture movement from the four perspectives: (a) communal self-help activities, (b) organizational structure, (c) linkages with a nationwide network of activists, and (d) marketing beyond the official certification systems.

5.6.1 Long-standing experience of community development

The establishment of communal financial institutions has been one of the major strategies for achieving the economic self-reliance of rural people since the heyday period in the 1980s when the development monks led community development. Based on this tradition, in the context of its activities NCS still assumes the role of a financial institution for both members and non-members. It operates a community bank, playing the role of a savings bank for its members while at the same time helping NCS to obtain the necessary capital to run the rice mill. The bank was started in the early 1990s, and in 2011 276 members and staff of NCS had deposited their money in it and were enjoying an interest rate higher than that offered by other commercial banks (NCS, 2012). NCS accepts deposits not just from its members but also from as many as 138 community groups, including a self-help group for meeting funeral costs located in Yasothon Province.

NCS also runs a welfare fund, created in 1993, to apportion some of the profits from the rice business for securing the welfare of its members. This fund is used to help members who experience a domestic calamity, including fire, damage caused by violent storms, and the death of close family members. The fund has even been deployed to repair a public road in a village for the sake of its entire community.

On the other hand, Rice Fund Surin provides loans for those members who cannot afford to bear their production costs. During the 2011–2012 season 89 members borrowed money amounting to nearly 9,000 baht per person to cover such production costs and used these funds mainly to hire a combine harvester (Rice Fund Surin, 2012). In addition, Rice Fund Surin's sister organization, CAEF, has evolved a fund for promoting sustainable agriculture. Non-organic farmers can thereby obtain loans for the conversion to an organic farming system. CAEF also provides micro-lending services for landless villagers to enable them to purchase land for organic farming or to pay off mortgages on their land (Nuntiya and Thunya, 2011).

5.6.2 Decentralized and multi-layered organizational structure

As previously discussed, Rice Fund Surin's management is characterized by its close affiliation with its partner organization, CAEF on the one hand, and with localized sub-groups on the other. The primary role of CAEF is to provide services and assistance that Rice Fund Surin as a business enterprise cannot provide for its members. A key activity of CAEF is to promote sustainable farming practices through seminars and workshops for concerned farmers, both members and non-members, of Rice Fund Surin. CAEF's other activities include the promotion of local alternative markets (discussed further in Subsection 5.6.4), promotion of food waste recycling, conservation of native plant seeds, and environmental education and awareness campaigns (Nuntiya and Thunya, 2011). As a powerful advocacy group, CAEF coordinates a campaign with sub-districts to promote sustainable agriculture.

Another organizational characteristic of Rice Fund Surin is that it is a loose federation of different sub-groups of local farmers scattered across Surin Province. This organizational structure contrasts with the centralized and unitary structure of the post-ACT Phathana Enterprise, the membership of which is confined to a specific sub-district (see Subsection 5.3.2.3). The sub-groups of Rice Fund Surin are not merely subordinated local chapters; rather, they are independent bodies. Although this independence of sub-groups has enabled some of them to become more business oriented (Section 5.5), it has also enabled each of the sub-groups to conduct activities oriented toward its local needs at the grass-roots level. Sub-groups established during the pre-ACT period (A, B, and C in Table 5.5) may represent the old guard of Rice Fund Surin. These sub-groups all include the words "natural farming" in their names, reflecting the nature of their organizational foundation grounded in the original alternative agriculture move-ment. The sub-groups' activities also reflect the basic attributes of the move-ment. A noteworthy example is a community seed bank created by sub-group A (see Table 5.5), with the goal of achieving self-reliance in relation to rice seeds. A number of members produce seeds of different varieties of jasmine rice, as well as an indigenous variety, by carefully selecting pure botanical lines. The majority of the seeds produced are sold to Rice Fund Surin, while the remaining quantity is basically sold within the group to its members.

5.6.3 Continued affiliation with the AAN

The continued association of Rice Fund Surin and NCS with the nationwide AAN network has contributed to the enduring commitment of farmers' groups to activities grounded in the original movement centering on alternative agriculture.

The linkage with the AAN has played a particularly significant role in preserv-ing local rice seeds, which the AAN, together with other influential NGOs, has worked toward in order to promote sustainable farming from the late 1990s and onward. In 2000 AAN activists conducted a field survey in northeastern prov-inces, including Surin and Yasothon, of local rice varieties that are on the verge of extinction. Subsequently, the AAN held seminars and training sessions for local farmers, focusing on the selection, breeding, and conservation of seeds, and facilitated exchanges of local varieties of seeds and strains between farmers (East Asia Rice Working Group, 2006).

A member of sub-group A of Rice Fund Surin (see Table 5.5), who is also an AAN activist, along with two other members of the sub-group, have grown a number of rice varieties in their own paddy fields, or in the sub-group's common field, since 2004. Having specialized in seed preservation, members of sub-group A have also embarked on seed production of different varieties of jasmine rice as another form of business. Such connections with the AAN were not evident among business-oriented sub-groups such as X and Y (see Section 5.5).[17]

5.6.4 New trends beyond official certification systems

5.6.4.1 Domestic markets for organic products without certification

Some organic farmers now prefer to sell their rice to domestic markets rather than to export it to foreign markets. Domestic organic businesses do not necessarily require third-party inspections based on the rigorous standards that are charac-teristic of international certification schemes. Regulations pertaining to the local organic market, if any, are more relaxed and flexible in nature, and producers can wield considerable discretion in setting up their own standards that are more fit-ted to local contexts and values. It is therefore not surprising that some organic farmers' groups have opted out of the thriving export business to sell their prod-ucts locally, thereby avoiding the laborious certification procedure.

A sub-group of Rice Fund Surin (not included in Table 5.5) presents a typi-cal example of such groups. From the outset, this sub-group has contributed to the establishment of the cooperative. When a foreign trader urged Rice Fund Surin to comply with EU organic standards for rice exports in 1999, this sub-group refused, claiming that Thai farmers should not be subjugated to externally imposed standards that were unsuited to their vision and practices. The sub-group subsequently left Rice Fund Surin and turned to the domestic market. When they sold their organic rice, in place of organic certification they used the name of the sub-group, *Sahatham* (united *dhamma*), to guarantee the quality of their products (Nuntana, 2001).[18]

5.6.4.2 *Emergence of local "green markets" for various organic products*

CAEF, the sister NGO of Rice Fund Surin, has actively promoted organic agricul-ture by linking producers directly with consumers. A striking outcome of CAEF's efforts has been the establishment of a weekly organic market that is regularly held in local towns. The number of such "green markets" has steadily increased from just one in 2003 to 14 across the provinces of Surin and Yasothon in 2014.

The Saturday green market held in Surin City is the oldest and largest of these markets, continuing to attract a growing number of organic farmers and consum-ers for more than a decade. The market was launched in 2003 to promote sales of various organic products as well as to generate regular incomes for farmers. The majority of participants in this market are organic rice growers associated with Rice Fund Surin, whose entire areas of farmland have already been officially certi-fied as organic. The number of producers supplying organic products to the Sat-urday green market increased from the initial 17 households to 86 households, being roughly equivalent to a quarter of the members of Rice Fund Surin in 2011 (Triiyadaa, 2012). The market is managed by the producer-farmers themselves, who also serve as organic inspectors in charge of monitoring the production pro-cess carried out by members, with active support provided by CAEF. The success of the green market in Surin City has prompted the establishment of a number of similar markets within and outside of the province (Tsuruta and Suriya, 2016). NCS also initiated its first green market in the provincial capital city of Yasothon in 2008.

Rice Fund Surin encourages its members to participate in the green markets to enable them to diversify their income sources and to avoid dependence on paddy sales. In the green markets, registered producers can bring anything they have and sell them directly to consumers. An astonishing variety of organic products are available: garden-fresh vegetables and fruits; wild herbs and berries; jasmine and traditional rice; organic pork and chickens; wild and farmed fish; processed foods, including fried crickets; and traditional local sweets (Tsuruta and Suriya, 2016). This demonstrates the fact that a number of members of Rice Fund Surin still practice integrated farming, which was initially promoted by the AAN as a way of diversifying farm outputs, in conjunction with commercial organic rice production. According to a survey conducted by the second author in 2013, 11 of 26 sample members (42 percent) of Rice Fund Surin who took part in the green markets earned larger annual incomes from the sales in green markets than from their organic paddy shipment for export.

Beyond the membership of Rice Fund Surin and the rice mill business, the green markets can encompass a wider range of farmers. Consequently, they offer regular income opportunities not only to ordinary farmers but also to disadvantaged villagers, who may not be able to produce organic jasmine rice for export. Among the products handled in the markets are many varieties of wild plants, fish, and insects that are easily collected or caught, even by landless farmers or elderly persons. CAEF especially encourages landless farmers to participate in the green markets (Nuntiya and Thunya, 2011; Triiyadaa, 2012).

CAEF made a further attempt to develop mutual solidarity between producers and consumers, which led to the Smiley Garbage Project in 2007. For this project organic farms use kitchen waste provided by urban consumers to make compost. By 2010 as many as 120 urban households, including regular customers of the Saturday green market, joined the project. Food waste from urban families was recycled to enrich the soil of 50 organic farmers (Thunya, 2010). Simultaneously, CAEF has run a campaign targeting consumers and emphasizing the importance of "eating local and eating organic." The green market event in Surin, which attracts 300 to 400 customers on each occasion, is a suitable venue for the CAEF staff to disseminate the messages of the campaign and to periodically organize relevant events. In addition, CAEF organizes study tours for urban consumers to visit organic farms (Thunya, 2010; Triiyadaa, 2012).

5.6.4.3 Resurgence of Buddhist values in organic production

As revealed in the historical background of Rice Fund Surin and NCS (Subsection 5.3.2), Buddhist faith may have contributed to the ongoing ethical initiatives of members of the pre-ACT farmers' groups. According to Kaufman (2012) and Kaufman and Mock (2014), many organic farmers in Yasothon Province still associate organic farming with Buddhist ethics, particularly the First Precept of not killing living beings. Although it is difficult to measure the intensity of the religious faith of organic farmers, a special project exists that is a manifestation of the enduring Buddhist ethics among organic farmers, as well as an offshoot of NCS.

This project, titled "Moral Rice," was originally conceived by a leading farmer (and also one of the founders of NCS) who wanted to inculcate moral responsibility among organic rice growers based on their religious status as Buddhists. This was an independent project that was not affiliated with NCS. The leader claimed that organic farmers must abstain from drinking alcohol, smoking, and gambling, as these behaviors not only went against Buddhist teachings but also contributed to the worsening debt problem of farmers. To embody this morality he established a new organization based in a temple near his home village and began to recruit like-minded "moral" farmers. The abbot of this temple was affiliated with a reformist Buddhist movement known as Santi Asok, which since the mid-1970s has been renowned for its vision and practice in pursuit of an ascetic and self-sufficient life based on vegetarianism and organic farming (Parnwell and Seeger, 2008).

The Moral Rice project gradually took shape in the mid-2000s and rapidly attracted participants during the post-ACT period (Table 5.2). Participants in the project, considered as "moral rice" producers, have to (a) abstain from all of the above-mentioned habits and (b) follow the Five Precepts of lay Buddhist practice (see note 4). Initially there were about 100 participant producers, 10 percent or less of whom were members of NCS. These uniquely Buddhist standards were designed not only to practice Buddhist ethics but also "to compete with external standards and expand the domestic market for organic produce" (Kaufman and Mock, 2014, p. 885). In 2006 rice produced under this project was certified as organic by ACT. While only 38 out of 108 registered participants met the Moral Rice project standards during the crop year 2006–2007, 119 farmers fulfilled the standards in 2012–2013, when organic paddy fields covered a total of 1,970 rai (about 315 hectares) of land (Moral Rice, 2014). The Moral Rice project has not only spread within Yasothon Province but has also gradually penetrated several neighboring provinces. The project also entails conservation of indigenous rice varieties. More than 200 indigenous varieties of rice are preserved in situ, in the farms of project participants.

Organic rice from the Moral Rice project is sold directly to urban consumers as well as through the facilitation of the Santi Asok movement network. More than 100 urban consumers support Moral Rice producers by paying them in advance before the harvesting season. This relationship appears to be similar to that of an ideal Fair Trade producer–buyer relationship. This kind of producer–consumer relationship has been established to remind urban consumers of their debt of gratitude to rice farmers and rice itself, which is not mere stuff to fill one's stomach (TV Burabha, 2010).

5.7 Concluding remarks

This chapter has shown how organic and Fair Trade initiatives in Thailand are based on homegrown social movements aimed at ameliorating economic and environmental injustice in rural areas. In doing so, it has focused specifically on three organic rice farmers' groups in the northeast. The AAN, Buddhist thinkers, development monks, NGO workers, and farmers have made concerted efforts toward their long-term goal of achieving both self-reliance of rural residents and ecological harmony. This goal is an ideal underpinned by the prevailing Buddhist ideology, which is epitomized in the core concept of *dhamma*. Because this concept embraces essential elements of Fair Trade and organic movements – social equity and environmental integrity – Thai activists and farmers have easily accepted these Western-derived initiatives.

However, the introduction of organic and Fairtrade certification systems has contributed significantly to the thriving rice export business. The original visions of a core social movement, the "alternative agriculture" movement – self-reliant and ecological communities on the basis of *dhamma* – were gradually replaced by a standards-based and export-oriented organic jasmine rice monoculture.

Nevertheless, at the same time, even after Fairtrade–organic rice production was established as a lucrative business, the movement-derived farmers' groups (Rice Fund Surin and NCS) have retained their original visions and aspirations centering on alternative agriculture, and have continually worked as community-centered organizations in tandem with their rice-exporting businesses. This could be attributed to their organizational structures that enabled the farmers' groups to work at decentralized sub-group levels and to receive assistance from non-profit organizations, broader networks, and urban consumers. This multi-layered and multi-dimensional organizational structure has allowed member-farmers to benefit from various activities other than the export of organic rice. In particular, the clearly delineated roles of Rice Fund Surin and CAEF have been effective: whereas the former can concentrate on profit-generating production and sales of organic rice, the latter is devoted to the promotion of sustainable farming practices and alternative markets at the grassroots level. In other words, these two organizations have complemented each other, working as a dual engine to attain common goals of farmers' self-reliance and protection of the rural environment. At an ideological level, enduring Buddhist values may have facilitated the retention of the visions of the original movement and ethical conduct among organic farmers.

Lastly, it is worth emphasizing that the alternative agriculture movement in Thailand has evolved toward the creation of domestic market channels for various organic products. There is a recognizable move away from commercial rice monoculture regulated by international certification schemes toward sales of diversified and self-certified organic products. The green markets located in the provinces of Surin and Yasothon have achieved significant success and contributed to the stabilization of farmers' incomes as well as the development of producer–consumer partnerships in support of the organic movement. The Moral Rice project, a more radical scheme to promote cooperation between producers and consumers based on Buddhist principles, is not merely an ideological backlash. Rather, it is an attempt to create a unique local brand that strongly appeals to Thai consumers. These new sales channels will help local farmers to reduce their dependence on precarious international markets for certified organic jasmine rice and could lead to the establishment of a community-based, self-reliant food system.[19] This direction may reflect the original principles of both the Fair Trade and organic movements.

Notes

1 Some labels were introduced not with the purpose of strictly guaranteeing organics, but to indicate the safety of foods and good agricultural practices entailed in their production (Roitner-Schobesberger, Darnhofer, Somsook, and Vogl, 2008).
2 Green Net Cooperative, which acquired its current name in 2001, was initially called Nature Food Cooperative.
3 "Socially engaged Buddhism" is a grassroots social movement led by intellectuals, NGOs, monks, and citizens within Asian Buddhist countries such as Thailand, Sri Lanka, and Cambodia. While criticizing Western development models focusing on economic growth, this movement has, since the 1980s, sought alternative ways of achieving social development in line with Buddhism, with the ultimate aim of building a harmonious society based on mutualism and a balanced ecology (Nishikawa and Noda, 2001, p. 15).
4 The Five Precepts comprise the following ethical principles, entailing abstention from (a) killing living things, (b) stealing, (c) engaging in sexual misconduct, (d) telling a lie, and (e) consuming alcohol (Ishii, 1991, p. 111).
5 A rice bank is a system wherein villagers deposit surplus paddy and borrow some, when necessary, at an interest rate that is below the commercial rate (Izumi, 1995).
6 Although there are no official data on the volume of Fairtrade products exported from Thailand, the number of Fairtrade producer groups listed on the FLO-Cert website indicates that rice is by far the most important Fairtrade item. Out of 32 Fairtrade-certified Thai producer groups, 15 groups deal primarily with rice (mostly jasmine rice), followed by 8 groups that deal mainly with fruits (FLO-CERT, 2016).
7 This Fair Trade project (discussed in detail in Subsection 5.3.2.1) was implemented before the acquisition of Fairtrade certification.
8 These organic certifications are verified by BCS Öko-Garantie, a German certification body. Rice Fund Surin considers these forms of certification to be more effective than the IFOAM-accredited certification, ACT, for export.
9 Twenty-nine percent of the funds that were ultimately raised comprised shares bought by urban consumers belonging to the association, and 57 percent comprised an advance paid for rice NCS produced (Nuntiya, 1998; Walaiporn et al., 2007).
10 The administrative structure in Thailand consists of the following units: province, district, sub-district, and village.
11 The World Intellectual Property Organization (2016) describes a geographical indication (GI) as a product sign that conveys a specific geographical origin and the product's possession of qualities or a reputation attributed to that origin. To function as a GI a sign must identify the origin of a product as a given place.
12 The duration of the crop cultivation season is from May to January.
13 After Prime Minister Yingluck Shinawatra was ousted as a result of a coup that took place in 2014, the farm-gate price of paddy returned to a normal level.
14 In 2012, 1 baht was equivalent to US$ 0.03.
15 During this season sub-group X may have accrued a large profit by selling its product through their own channels. Rice Fund Surin purchased organic paddy from its members at approximately 20 baht per kg on average, while milled organic rice was sold at 50 baht per kg. Based on the assumption that the weight of milled rice is half that of the paddy prior to milling, it is estimated that milled rice can earn a profit that is 25 percent higher than the value of the unprocessed paddy.
16 NCS has also reportedly experienced a similar tension. In the mid-2000s a conflict occurred among its leaders over the setting of farm-gate prices because some members wanted a more business-oriented management (Parnwell, 2005).

17 A similar story has unfolded within NCS. In 2005 one of the leaders of NCS, who was also an AAN activist, organized a sub-group that specialized in the preservation of traditional rice varieties.

18 A similar pattern of localism and rejection of export-oriented certification schemes has been observed in Chiang Mai in northern Thailand. A local certifying agency, the Northern Organic Standards Organization, has attempted to create a locally embedded certification system based on local values and community recognition rather than on rule-bound, universal validation and labeling schemes that tend to benefit only large-scale farmers (Vandergeest, 2009; Wyatt, 2010).

19 Somporn and Fukui (2005) argued that Thailand is less competitive in the global jasmine rice market than are neighboring countries that joined the market later, namely Cambodia and Laos.

6 Certification-supported farming and other diversified livelihoods

A case from the Philippines

Although certification is situated inside the farm sector in the previous chapters, in reality certified farmer groups are involved not only in farming but also in non-farm activities in parallel. In this chapter we attempt to explore the convergence of Fair Trade and organic initiatives from a wider perspective of rural development by focusing on plausible tensions between certification-supported farming and other diversified economic activities.

6.1 Relationships between agricultural certification and diversification

Debates about the role that certifications such as Fairtrade and organic can play in the livelihoods of farmers in the global South have tended to focus on the direct impacts of certification, such as price premiums and trading relationships with buyers (e.g., Bacon, 2005; Bolwig, et al., 2009; Jaffee, 2007; Nigh, 1997; Ruben, 2008). As a result, little or no attention has been given to the critical issue of how certification interacts with structural agrarian changes underway in the rural South. Although such certification systems contribute to agricultural development in the global South, this approach may be in conflict with what is broadly recognized as the significant expansion of non-farm activities in the rural economy (e.g., Bhaumik, 2007; Buchenrieder, 2005; Goldstein, Childs, and Wang-dui, 2008; Lindberg, 2012; Rigg, 2006). According to Haggblade, Hazell, and Readon (2010, p. 1429), non-farm earnings currently comprise 30–50 percent of rural household incomes in the developing world, and this share is increasing. Barham, Callenes, Gitter, Lewis, and Weber's (2011) study of Mexican coffee producers demonstrated that more income is derived from off-farm labor opportunities than from certified or non-certified coffee cultivation. Poor farmers have to diversify income sources both inside and outside the agricultural sector as a coping strategy (Ellis, 2000, pp. 61–66). If the shift from farm to non-farm activities is a core part of broader development processes, agricultural development with certification may not be an adequate policy. As long as agricultural certification is used for Southern producers, its interaction with ongoing agrarian changes must be considered in order to understand the realities of certification.

Some authors have discussed relationships between agricultural certification and diversification mainly as a theoretical critique of the Fair Trade system and counterarguments to it, although they do not clearly define diversification. LeClair (2002) argues that the "artificial" increase in the price of certified products prolongs dependence on the products with poor long-term economic prospects and "retards the diversification of production that is necessary for the economic advancement of developing countries" (p. 957). Along the same lines, Sidwell (2008, pp. 13–14) criticizes Fair Trade for assuming "that poor farmers must always remain farmers" and for "deny[ing] the possibility of diversification." Although organic certification is not similarly criticized, the commercial monoculture production of organic cash crops is regarded as a potential problem of organic agriculture (La Trobe and Acott, 2000, p. 314). In contrast, Hayes (2008) and Smith (2009) contend that the perceived options of farm and non-farm activities are false. Instead, they argue that Fair Trade offers the additional inputs of economic, social, and physical capital (Hayes, 2008), thereby "alleviating the capacity constraints that normally hold back diversification by the poor" (Smith, 2009, p. 470). To simplify these arguments, certification is interpreted as the sole factor directly influencing diversification. However, in most real settings factors other than certification simultaneously shape and influence diversification, and it is impossible to clearly differentiate the impact of certification from that of other factors. In some cases agricultural certification may be introduced to households that have already diversified their livelihoods; diversified activities also influence certification-supported farming. Therefore, it is important first to understand how producers belonging to a certified cooperative diversify their livelihoods and then clarify the role of certification within the producers' overall livelihood strategies. The objective of this chapter is not to analyze the causal relationship between certification and diversification, but to situate diversification as a given context and to explore the meaning of agricultural certification in a rural setting where diversification is underway.

The next section first defines the concept of diversification for the present research and proposes a framework for interpreting diversification in real rural settings. The examination of diversification and other theoretical propositions then raises questions to be addressed in a case study. The research approach is outlined in the third and fourth sections. The case study adopted is a Fairtrade and organic-certified sugarcane cooperative in the Philippines, where land reform remains an important agrarian issue (see World Bank, 2009). The cooperative was organized to help land reform beneficiaries – formerly landless plantation workers – obtain farmland. Although land reform beneficiaries are expected to be eager to take up farming as new landowners, the Philippines is also known for its highly diversified rural communities, with a high incidence of emigration (see Ang, Sugiyarto, and Jha, 2009).[1] In this setting a variety of diversification can be observed explicitly. Using the framework proposed in the second section, the sixth section interprets the empirical observations of diversification practices underway in the community. The seventh section then analyzes why and how certification-supported farming coexists with other diversified activities. The chapter concludes by summarizing the major findings from the case study and suggesting, as policy implications, conditions for the effective use of Fair Trade and organic initiatives with livelihood diversification.

6.2 Theoretical propositions

6.2.1 Definition of diversification

As Hayes (2008, p. 2598) suggests, "the question of diversification looks different from the perspectives, respectively, of the individual household and of the community or country as a whole." When diversification is observed at the community or country level, it appears as structural agrarian change. In this case study diversification is observed at the household level, namely livelihood diversification. Adopting the definition of Ellis (2000, p. 15), this chapter interprets rural livelihood diversification as "the process by which rural households construct an increasingly diverse portfolio of activities and assets in order to survive and to improve their standard of living."[2] Livelihood diversification at the household level can be practiced as multiple activities of an individual family member, different activities of multiple members, or both. The scope of this research uses households as the analytical unit, which restricts it from shedding light on gender and generational differences inside households.

Although a shift from farming to non-farming activities is observed as structural agrarian change (see Section 6.1), livelihood diversification at the household level does not always move in the same direction. Livelihood diversification may lead individuals away from farming or into farming; remarkably, the latter case has been observed among landless people (Hossain, 2004, p. 4054). Important diversification also occurs within the crop portfolio of a given farming household. Furthermore, the space for livelihood diversification has already expanded beyond the village (Wilson and Rigg, 2003, p. 696). This chapter assumes that livelihood diversification can occur both away from and into farming, and both inside and outside the original rural setting.

6.2.2 A framework for interpreting livelihood diversification

The literature concerned with diversification has also focused on the motivations of actors. In general, diversification is connected with two contrasting motivations: survival or accumulation (Hart, 1994). These are sometimes differentiated as necessity or choice (Ellis, 1998). More specifically, Scoones (1998, p. 9) suggests that "diversification aimed at coping with temporary adversity" is a typical strategy chosen by poor people, whereas an "active choice to invest in diversification for accumulation and reinvestment" is usually practiced by wealthier actors. Complementing this dichotomy, Makita (2007, p. 68) argues that "if actual livelihood diversification is an outcome of balancing the two types [survival and accumulation], there should be a shifting process from diversification for coping with temporary adversity to diversification for accumulation and reinvestment." For example, in rural Bangladesh Makita (2007) identifies three phases of livelihood diversification undertaken by landless households: for survival (Phase 1), from survival to accumulation (Phase 2), and for accumulation (Phase 3), and in this context the author interprets Phase 2 as a pathway out of poverty at the household level. Although the specific activities undertaken in each phase and the factors determining each phase are likely to vary by context, these three phases of livelihood diversification can apply as an interpretive framework for any rural setting.

In interpreting livelihood diversification underway in the case study setting, this chapter draws on this three-phase framework. Theoretically, diversification can be analyzed from two perspectives: long-term dynamics and short-term statics. Although the former perspective reveals how, since its introduction, certification has contributed to its beneficiaries' long-term upward mobility from one phase to the next, it is physically impossible for outsiders either to identify the starting point of each household's diversification or to observe the household continuously. Instead, adopting the second perspective, a one-time case study, illuminates which phase each household currently belongs to as a result of both certification-supported farming and diversification. In brief, diversification in this study is analyzed not as a *process* but as an *outcome*. The present study initially observes how the three phases of livelihood diversification are delineated, identifying factors that signify the respective phases, and then identifies the phase of livelihood diversification each household has reached.

6.2.3 From certification to certification-supported farming

There are at least two reasons why the direct causal relationships between certification and diversification cannot be empirically analyzed at the household level (see Section 6.1 in this chapter). First, certification interacts with a certified producer cooperative, not with individual members. When agricultural certification is introduced to a small farmer cooperative, all activities related to certification, including training for farmers and the use of Fairtrade premiums, are conducted through the cooperative (e.g., see Utting, 2009). These activities are also technically and financially supported by intermediary organizations (e.g., Schmitt, 2012; Tallontire, 2000). Therefore, cooperative members benefit from certification only through the cooperative system. Second, no one differentiates the financial benefit of certification, such as price premiums, from other financial sources when investing in a diversified activity. Even if an increase in income that certification has brought motivates diversification, the new activity itself may be financed by a loan from the cooperative or profits from another previously diversified activity. We can only assume that certification influences diversification *indirectly*. For these reasons, in this chapter certification is studied as certification-supported farming conducted through the cooperative system.

6.2.4 Research questions for the case study

To provide an empirical contribution to the theoretical debate on the relationships between agricultural certification and diversification, this chapter addresses two questions: (a) how small farmers belonging to a certified cooperative diversify their livelihoods and (b) why and how their diversified activities are or are not made compatible with certification-supported farming.

The focus on the compatibility of diversified activities with certification-supported farming derives from tensions that can occur physically between farming and non-farming livelihoods. As Rigg (2005, 2006) argues, diversification away from farming has become a key factor in determining the levels of income and well-being for rural households in many parts of the South. Assuming that policy makers consider the rural non-farm economy as a potential pathway out of poverty, Haggblade, Hazell, and Readon (2002, 2010) propose greater investment in the rural non-farm economy and increased access for the poor to growing market niches. Increasing non-farm activities is likely to conflict with certification-supported farming. For example, Lewis and Runsten (2008) report that increasing migration – a form of diversification – makes it difficult to continue Fair Trade coffee production in Mexico.

The same type of tensions may emerge more apparently between diversified activities and sustainable farming practices that both Fair Trade and organic certification systems demand. Fair Trade standards, albeit less strict than organic standards, require participant producers to conserve natural resources through sustainable agricultural practices (see FLO, 2011c). The question emerges as to whether producers can continue required sustainable practices in the face of diversification away from farming. The existing literature suggests that diversification by poor farmers into non-farm activities generally heightens unsustainable farming practices. Based on the analysis of household-level data collected in nine regions of seven countries in Africa and Asia, Kuiper, Meijerink, and Eaton (2008, p. 90) conclude that "income from non-farm activities is neither invested in agriculture, nor in ensuring future production." Sustainable farming practices cannot be expected in such a situation. In the Philippines, Rola and Coxhead (2002) and Parrilla (cited in Lumley, 2002, p. 89) each reported reduced attention to soil conservation practices on farms as a negative effect of off-farm employment, at least in the short term. In an attempt to introduce organic farming to a farmers' association in South India, it was particularly difficult to convert association members with non-farm incomes from conventional to organic farming because they were reluctant to give more time to farming (see Chapter 2). The results of an empirical study conducted in Honduras by Morera and Gladwin (2006, p. 375) also indicate that "off-farm work competes with soil conservation practice because . . . rather than use their off-farm income to boost their farming with improved technologies, farmers tend to relegate their farming to food security only, reducing the labor applied to it." How can the negative influence of diversification be removed? The two research questions are explored here using the case of a certified sugarcane producer cooperative in the Philippines.

6.3 The setting: Land reform in the Philippines

Although the trend of farmers diversifying away from farming is remarkable, in certain contexts there are large categories of landless workers who aspire to become landowning farmers.[3] In the case of the Philippines, the desire is motivated by the strong association between inequality of landownership and rural poverty (Asian Development Bank [ADB], 2005, pp. 64–65; 2009, pp. 46–47). Although since the 1930s several land reform programs have attempted to redress this (Reyes, 2002, pp. 6–11), significant inequality in land distribution still existed in 1988: 60 percent of farming families were landless, while fewer than 0.1 percent of families owned more than 25 percent of all landholdings of 100 hectares or more (Pye-Smith, 1997, p. 7). For this reason, the government launched the Comprehensive Agrarian Reform Program (CARP), the latest program to redistribute public and private agricultural land to tenant farmers and farm workers. Under CARP, large private landowners were expected to sell their land to the government or directly to peasants, who were expected to buy land with a government-supported loan to be repaid over a 30-year period (Diprose and McGregor, 2009, p. 52). Whereas earlier land reform programs were primarily limited to paddy and corn fields, CARP was the first program to cover sugarcane lands (Billing, 1993, p. 130). An opportunity to own a plot of farmland was thus finally extended to landless sugarcane hacienda (plantation) workers (Billing, 1993, p. 130; Provincial Government of Negros Occidental, 2007, p. 11).

Sugarcane has been the country's largest non-cereal crop in terms of planted area and production value, playing an important role in the Philippine economy (World Bank, 2009, pp. 24–25). Under the Spanish rule, by 1841 sugar was already the colony's leading export (Billing, 1993, p. 124). The establishment of the sugar industry created "an indigenous planter class" and formed a "society polarized between the country's wealthiest regional elite and its most volatile working class" (McCoy, 1992, pp. 107, 137). Filipino sugar planters have continued "to exercise extraordinary political influence" also "in the emergence of the modern Filipino nation state" (McCoy, 1992, p. 107). The Philippine hacienda was not only a farm but also "a residential unit for workers and their families, containing shops, schools and churches" (Billing, 1993, p. 130). For workers whose lives had depended on haciendas for generations, CARP meant more than just land redistribution.

Although CARP has been evaluated as comparatively successful (Borras, 2001; Borras, Carranza, and Franco, 2007; Reyes, 2002), it has not been easy for sugarcane plantation workers to acquire farmland. As expected, in the face of plantation owners' strong resistance the government was compelled to assign low priority to the redistribution of sugarcane lands (Billing, 1993, p. 130; World Bank, 2009, p. 25). Indeed, even after acquiring land successfully the recipients tended to cultivate sugarcane on their own without training, extension services, credit, or marketing assistance, and many ended up reselling the land, concluding that they could not independently manage viable farming (Billing, 1993, p. 130). Although the state and labor unions promoted the formation of cooperatives for land reform beneficiaries, "most cooperatives died a quick death after beneficiaries pulled out to start on their own, once they were allotted their individual plots. . . . Many are now leasing back their lands to planters or to small entrepreneurs, working as wage laborers on their own land"

(Rutten, 2010, p. 212). Many struggles have also been reported, not only between planters and activist workers (as potential beneficiaries), but also among the landless workers themselves over access to land (Rutten, 2010).

To improve land reform outcomes, some NGOs have worked to facilitate the transfer of land from plantation owners to landless workers. NGOs therefore tend to be described as promoters of social movements (e.g., Franco, 2008; Rutten, 2010) because their activities have focused on increasing the unity of potential beneficiaries, their education, and collective negotiation with the government and other related stakeholders (Borras, 2001; Borras and Franco, 2007; Diprose and McGregor, 2009; Liamzon, 1996). By the late 1990s, however, many NGOs had shifted the focus of their development work from support for land redistribution to "farm development," although not all land transfers had been completed successfully (Borras and Franco, 2007, p. 24).[4]

The case study analyzed in this chapter deals with Fairtrade and organic certifications used for such a program of "farm development" by former sugarcane plantation workers in the province of Negros Occidental. The island of Negros, composed of Occidental and Oriental, possesses the most inequitable distribution of land in the Philippines; when CARP began, more than half the available agricultural lowlands were under sugarcane monoculture on large plantations, producing about two-thirds of the country's total crop (Diprose and McGregor, 2009, p. 56). A local NGO named Alter Trade Foundation Incorporation (ATFI) has since 1995 provided technical assistance and production loans to support associations of former plantation workers who are land reform beneficiaries. Furthermore, an associated company, Alter Trade Corporation (ATC), has purchased sugarcane from these associations and exported traditional brown *muscovado* sugar to Fairtrade and non-certified ethical markets in wealthy consumer countries in Europe and Northeast Asia.[5] Although many cooperatives formed by land reform beneficiaries were eventually disbanded under pressure from members for individual landownership (Rutten, 2010, p. 212), those under the auspices of ATFI and ATC have continued to farm sugarcane collectively after the redistribution of land to maintain the economies of scale in former plantations.[6] According to ATFI and ATC staff members, Fairtrade and organic certifications were introduced to increase the returns from sugarcane production and to facilitate the repayment of government-supported loans used to acquire the land.

6.4 Outline of the case study

6.4.1 The case: A cooperative of land reform beneficiaries in Negros Occidental

Empirical research focused on the oldest of 14 former workers' associations supported by ATFI (as of August 2011). The association, currently called Santa Rita Farmers Multipurpose Cooperative (STARFA),[7] was organized in Purok Santa Rita, Barangay Dulao, in 1994.[8] Since that time the membership had increased from 60 to 73 members without any of the original group having left the association. One household was considered the equivalent of one member. STARFA had leased 20 hectares of sugarcane fields from a plantation owner for the past 17 years. Although the members were scheduled to take *collective* ownership of the 20 hectares soon, they had not yet received an official landownership certificate issued under CARP.[9] They had paid and had to continue to pay the lease fee to the plantation owner as long as the issue of a certificate was suspended. The assistance provided by ATFI and ATC for STARFA began in 1995 and was for the promotion of organic cultivation, the purchase of sugarcane, and financial support. All 14 former workers' associations, including STARFA, were certified as organic in 1995 and as Fairtrade in 2004, and their organic sugar had been exported through ATC to Northern ethical markets since 1995. STARFA was officially registered as a cooperative in 2007.

A preliminary visit to ATFI revealed that the cooperative differed from the other ATC-affiliated producer associations in three additional respects. First, the cooperative had not yet acquired a collective ownership certificate, although the other 13 associations had already done so. However, this was not necessarily a special case. By 2005 only 61 percent of targeted land in the province had been redistributed (Tecson, 2006, cited in Rutten, 2010, p. 207). Second, many members of the cooperative expected to obtain an *individual* ownership certificate for other plots of farmland separate from the communal land, whereas the other associations only operated collectively. When the cooperative gained the communal 20 hectares for its 60 members (0.33 hectares per person), the land was too small for all the members to rely on it as their main income source.[10] Those who could afford to access land individually had been waiting for both collective and individual ownership certificates. Third, their community was closest to the capital city of the province, Bacolod. This cooperative may therefore have facilitated our observation of greater diversity in land use and livelihoods than was the case in the other associations.[11]

6.4.2 Data collection

Primary data were collected by the first author in Purok Santa Rita and Bacolod City in July and August 2011. Fifty-eight of 73 cooperative member households lived in Santa Rita, and semi-structured interviews were carried out with 53 members (3 members were too busy, and 2 were unavailable due to illness or disability). Although access to all adult members of each household was attempted, such full interviews were realized only with 10 households. Standard questions for each member household focused on how the household was involved in the cooperative, what income sources were available inside and outside the cooperative, and what the perceived benefits were from membership. As explained in the next section, the remaining 15 member households had migrated away from Santa Rita, although they retained membership in the cooperative. The same type of interview was conducted with one person from this group during a return visit to attend a monthly meeting of the cooperative. Information on the operation of the cooperative was collected mainly from current and former board members within the sample of 53 households. Neither the cooperative nor its members consented to disclose their detailed financial information, which restricted the analysis to qualitative data only. As an alternative, members themselves identified as economically successful or economically unsuccessful.

To triangulate data collected from the cooperative members, supplementary interviews were conducted with four ATFI staff members and two members from ATC; two board members of the umbrella organization, Negros Organic Fair Trade Association (NOFTA), which comprised all fourteen associations; three former plantation workers who lived in Santa Rita but who did not belong to the cooperative; and the head of a monastery located in Santa Rita who had lived and worked there since 1991. In addition, a monthly meeting of the cooperative and a regular meeting of NOFTA were observed. One-day visits were also made to two of the other associations located farther away from Bacolod to gauge the representativeness of STARFA vis-à-vis the others in the wider organization.

6.5 Certification-supported farming in the case study setting

The primary objective of the cooperative was to enable all members to become landowners, and some also aimed to obtain individual ownership. All members understood that collective action through the cooperative was indispensable for gaining a land title. Members had continued, and intended to continue, to pay lease fees until they obtained landownership certificates, after which they were required to amortize acquisition fees. To become landowners they had to maximize profit from their land and gain additional income-generating opportunities. When farmland was distributed to former plantation workers, ATFI guided the associations to operate sugarcane fields collectively and to add organic value to their production because sugarcane production without disbanding the plantation made best use of the land. According to the managerial staff of ATFI there was no alternative cash crop equal to sugarcane, but little of the sugarcane produced in the Philippines was organic in the mid-1990s when ATFI introduced organic farming to the associations' communal lands. Certified organic sugarcane production had enabled ATC to export the association's produce to specialized Northern markets. Fairtrade certification had further increased ATC's access to export markets.

The ATFI-assisted associations did not encounter substantial challenges in converting from conventional to organic farming. In general, small farmers or landowners hesitate to switch from conventional to organic farming for fear of appreciable decreases in productivity (Eyhorn, 2007, pp. 141–144). However, these former plantation workers accepted organic practices easily. When plantations employed association members they were all ordered to use chemical fertilizers and pesticides, but they did not pay attention to yields on the plantations. Only after land was distributed to them did they begin sugarcane cultivation responsibly. Although there may have been a decrease in yields from the same plots of land, it was impossible for association members to compare the productivity of conventional farming on the plantations with organic farming on their own communal land. They simply appreciated the opportunity to export to Northern organic markets.

In STARFA one member was appointed as a farm manager, receiving a fixed salary from the cooperative. He planned the production schedule on the communal land for each year, and the board members approved it. The farm manager calculated the workload necessary for each season and decided the employment of workers for each day. Anyone from the member households was able to work and earn wages that were separate from dividends received equally by all the members. There was no difference in wage rate between male and female workers, but males tended to take on physically harder work with better wages. The farm manager took responsibility for employing workers from outside the cooperative when there were not enough applicants from the member households. He noted there was no difficulty in hiring workers from more remote areas of the island. The same wage was paid to member and non-member workers. The farm manager also arranged the shipment of sugarcane in response to orders from ATC.

Organic and Fairtrade certifications had contributed to the capacity of this sugarcane producer cooperative in various ways, although their direct monetary impact was not large. In principle, ATC purchased sugarcane only from the collectively operated fields of each association. The cooperative sold about 40 percent of its total communal production to ATC each year. Due to financial constraints (detailed later), at the time of the author's fieldwork, ATC's purchase prices were kept at the level of local conventional market prices; the cooperative did not receive organic price premiums from ATC. However, they had benefited from the minimum price guaranteed by Fairtrade certification, and they had used Fairtrade premiums to renovate their office building and to buy a computer. A portion of the premiums was also used for the production of organic fertilizers using *carabao* (buffalo) dung and other locally available manures.[12] This was thought to bring the dual benefits of cheaper production costs than the cost of purchasing chemical fertilizers and an additional income stream generated through the sale of organic fertilizers to outside farms. Its association with ATC also allowed the cooperative to access advantageous production loans from ATFI every year.

It was not possible to conduct research with a control group of non-member former plantation workers in the same locality. It was more difficult to get in contact with non-member former plantation workers because most of them worked outside Santa Rita. Hence, interviews were carried out with only three non-member households. Two of the three households depended primarily on income from *conventional* sugarcane cultivation on their individually acquired land under CARP, and the head of one household was regularly employed by a farm. Although the two households that cultivated sugarcane expressed their wish to join the cooperative, the cooperative was not able to accept new members due to ATC's financial constraints. According to the testimony of the head of the monastery – who had observed people in Santa Rita as an outsider for two decades – cooperative members were generally better off than non-members were.

This person witnessed many non-members subletting their individually acquired land because of financial constraints and becoming employed as laborers again. The cooperative's current chairperson also indicated that, as a result, neighboring non-members had diversified away from farming for survival more frequently than had cooperative members.

6.6 Livelihood diversification practiced by the cooperative member households

6.6.1 Overview from a long-term perspective

When the cooperative was organized, all the original members except two were landless plantation workers. Those two members, wealthier non-farm workers who had stable employment in the local government and a large company, respectively, were exceptionally asked to join the organization to enhance its financial capacity because it was particularly important to collect as much capital as possible in the initial stages. The economic status of the other members was almost equal, but over time a clear economic disparity emerged. Although the change from plantation workers to joint farm owners was regarded as an advance in livelihood, the members felt that such a way of living was insufficient to escape poverty. Fifteen members chose to migrate permanently out of the Barangay, abandoning agricultural life (Table 6.1). However, even such non-resident members, having at least paid capital shares, were counted as members and were able to receive dividends from the cooperative's profits; some made it a custom to return to Santa Rita once a year to receive these. Many members living in Santa Rita also leased farmland individually or had sent family members to other islands

Table 6.1 The largest income sources of STARFA member households (Negros, the Philippines)

Income source		No. of households
Off-farm outside the Barangay		15[a]
Inside the Barangay:		53[b]
On-farm	Wages and dividends from STARFA	(6)
	Individual farm	(30)
Off-farm[c]	Non-farm employment	(9)
	Remittances	(3)
	Salary from commercial farms	(2)
	Pension/ assistance from children	(3)

Source: Field survey

Notes:
[a]Fifteen of the 73 member households had migrated away for non-farm activities. Here Barangay, the second lowest administrative unit, is used to define permanent migration.
[b]Fifty-three of the 58 members living in Santa Rita responded to the survey.
[c]This study regards income from members' communal and individual land as *on-farm* and wage employment in agriculture earned on other people's farms as *off-farm* by adopting the concept of off-farm income from Haggblade et al. (2002, p. 3).

or abroad. In 2011, only six member households relied on the collectively operated land of the cooperative as their primary source of income. They made a living from wages for physical work on the communal land and a dividend from the profits of the cooperative without any individual farmlands or other stable employment opportunities.

Thirty member households had leased farmland ranging from 0.25 to 3.0 hectares for an individual ownership certificate in addition to the communal land.[13] For them, sugarcane production on their individual land was the largest source of income (Table 6.1). Although small-scale farming was generally not suitable for sugarcane cultivation, they asserted that the average profit from even the smallest individual plot (0.25 hectares) surpassed a member's dividend from the cooperative – the per-member profit from the communal land (0.27 hectares per person) after paying labor and other costs necessary for the operation of the cooperative. The cooperative members sold sugarcane from their individual land to other private mills that supplied to non-Fairtrade, non-organic markets. Although they did not receive any price premiums from their individual land, they also purchased organic fertilizers produced by the cooperative and practiced organic farming on their individual sugarcane fields, expecting to sell to ATC in the future.[14] Members with their individual land tended to give priority to farming these plots. Although some of these 30 members also worked for the communal farm when they had extra time, they were busy on their individual land in the harvest season from September to March. Consequently, harvesting on the cooperative's communal land was highly dependent on waged labor from outside. For members whose primary income was from their own individual land, the collective land operation did not seem to be a significant source of income.

The primary income source of the remaining member households living in Santa Rita was no longer on-farm activities (Table 6.1). Separately from the two members who had been non-farm workers since the beginning, seven members had switched from farm to non-farm occupations, such as *jeepney* (local minibus) or truck driver, security guard, office clerk, or shop manager after they had joined the cooperative. For three member households, remittances from abroad or the capital city, Manila, were their largest income sources. Additionally, there were two members who preferred to work regularly for commercial farms outside Santa Rita. Three members depended on pensions or assistance from others. All these 17 members kept up their memberships as a secondary or tertiary income source. In 23 of the 54 interviewed member households, females were registered as official members, mainly because male family heads and sons were too busy to attend the cooperative's monthly meetings. Indeed, the author's observation of a monthly meeting found that only 37 persons (one from each member household) participated, the majority of whom were female (Figure 6.1). Many member households did not seem to attach importance to income from the cooperative.

6.6.2 Interpretation of the current phases

How can the outcomes of livelihood diversification, noted previously, be interpreted using the three-phase framework (see Subsection 6.2.2)? All interviewed

Figure 6.1 A monthly meeting of STARFA (Negros, the Philippines)

members suggested that the six member households that depended on the cooperative as their primary income sources without individual farmland were the poorest in the community. For those six households all adult family members attempted to find as many jobs as possible in the cooperative, and the cooperative gave priority to those six member households in providing additional job opportunities. The households also attempted to complement the income from the cooperative by engaging in activities that required no capital, such as charcoal making, or they reduced household expenses by sending a family member to work outside the island. The head of one of the six households, JB, was a representative example. How he and his family made their living is described as follows:

> JB was born in Santa Rita and currently lives with his wife and three children. His eldest daughter works as a domestic worker in Manila. JB joined the cooperative in 1994. Three members of his family work for the cooperative: JB, his wife, and one son. They go to the cooperative office to ask about the availability of jobs every morning. Not all applicants can always find jobs in the cooperative during the agricultural lean seasons. For instance, in July 2011, whereas JB worked for 22 days and his son for 10 days, his wife was not able to find the lighter work she preferred. JB regularly takes charge of one of the three *carabaos* owned by the cooperative. With this *carabao*,

he ploughs the communal land as well as other farmlands, retaining 80 percent of the wage rate and handing over 20 percent to the cooperative. JB transports *carabao* dung for the cooperative to produce organic fertilizer. JB receives a bonus equivalent to one month's wages and microcredit as benefits from the cooperative. His family has used the microcredit mainly for school fees and other educational purposes. Although JB's daughter sometimes remits part of her salary from Manila, wages earned by the three from the cooperative are much larger than her remittances. The other two children have received small scholarships from the cooperative. One daughter has just graduated from high school, but the family cannot afford to send her to college. JB has wanted to lease a plot of farmland individually but has never had sufficient funds to do so.

The six member households, including JB's, can be regarded as still being in Phase 1 of livelihood diversification: they *must* diversify income-generating activities for survival.

The remaining 45 member households, who were wealthier than these 6 households, individually leased farmland for future ownership, regularly engaging in non-farm self-employment, or regularly receiving remittances from abroad. Although it is impossible to compare these households in terms of economic status without detailed financial data, the members themselves admitted that they were at different economic levels. Access to individual land in addition to the communal land was an important factor that had enabled them to invest part of their income in the diversification of their income sources. For example, the livelihood strategy of AC, the vice chairperson of the cooperative who was in his mid-30s, is described as follows:

> AC was born in Santa Rita and went to Manila after finishing high school. In 2008, he returned to his home village with his own family after his father's death and succeeded his father as a STARFA member. AC's father, one of the original cooperative members, had leased 1.2 hectares for a future individual certificate (1 hectare for sugarcane and 0.2 hectares for paddy) and AC has maintained the lease, although the family had to sell their *carabao* to afford medical treatment for his father. AC regularly works for the cooperative's communal land and also manages the leased land intended for his ownership. AC never works for other farms as an employed worker. His wife operates a *sari-sari* store (convenience store) at home, caring for their three small children. Thanks to microcredit from the cooperative, they can stock goods for the store without borrowing from moneylenders. In 2011, AC leased one additional hectare for three years to expand his sugarcane production.

AC's livelihood portfolio is beyond Phase 1 of diversification for survival, but it is difficult to assume that the purpose of the household's diversification is accumulation and reinvestment. AC was prepared to continue to work hard both inside and outside the cooperative in order to fund the education of his three children.

For these reasons, AC's way of livelihood diversification falls best into an intermediary stage, Phase 2, for the transition from survival to accumulation.

Although it is impossible to classify all the remaining member households clearly into either Phase 2 or 3, at least two households had steadily increased the number of their income sources, continuing to reinvest part of their income into the existing diversified economic activities on a larger scale than AC was doing. The cooperative members all agreed that a former chairperson of the cooperative, TT, was the most successful member among those who had started as a plantation worker in Santa Rita:

> TT, one of the original members, has *never* worked on the communal land. Fortunately, his father acquired a 4.5-hectare paddy field in the 1970s under a previous land reform program. Although TT himself worked as a plantation worker before he inherited the paddy field from his father, his income from the paddy enabled him to proceed with his diversification initiatives. At present, he leases three hectares for an individual ownership certificate and another five hectares from five people outside the cooperative who cannot finance the operation by themselves. TT supervises all these plots of land, employing workers throughout the year. He also owns two *carabaos*, a water-pump for irrigation, and a hand tractor. One of his daughters manages a retail shop, selling rice from his paddy field. His major non-farm income comes from a transportation business with his two trucks. The drivers are one of his sons and a son-in-law. Furthermore, he was elected and worked as a Barangay councilor for eight years until 2010. Even such a successful member uses microcredit to buy merchandise for his *sari-sari* store.

TT's livelihood diversification can be interpreted as falling into Phase 3, for accumulation. TT was fortunate not only in his individual access to farmland but also in other assets – at least his father's paddy field – that were available to him *before* he participated in the cooperative. Without those assets, TT would not have been able to reach such a higher economic status. Another member household that the cooperative community recognized as having a highly beneficial and diversified livelihood portfolio had also inherited paddy land. Except for the two members who had never farmed, there were a few more member households engaging exclusively in non-farm activities that the cooperative community recognized as wealthy. Although the diversified portfolio of these households might fall into Phase 3 of livelihood diversification, the assets they used for diversification were not disclosed. Such assets may have included personal networks that are indispensable for success in international migration (Tyner and Donaldson, 1999).

Table 6.2 summarizes how the situation of the cooperative member households' livelihood diversification at the time of the study can be interpreted through the three-phase framework. Whereas all members of the cooperative benefited equally from certification-supported farming, the differences in the level of livelihood diversification derive from factors other than the effect of certification. In the case of this cooperative, the determinants are whether a household is

Table 6.2 The phases of livelihood diversification reached by STARFA member households with the three-phase framework (Negros, the Philippines)

Livelihood diversification	Communal farmland (under STARFA)	Individual farmland	Other assets before participation in the cooperative	Involvement in non-farm activities	Economic status[a]	Estimated number of households[b]
Phase 1: For survival	Accessible	Not accessible	Not accessible	Part-time	Lower	6
Phase 2: For the transition from survival to accumulation	Accessible	Accessible or not necessary	Not accessible	Part-time or full-time	Middle	43
Phase 3: For accumulation	Accessible	Accessible or not necessary	Accessible	Part-time or full-time	Higher	2

Notes:
[a]The economic status was based on members' evaluations.
[b]Of 53 interviewed households who lived in Santa Rita, the 2 exceptional non-farm households were excluded.

able to obtain additional farmland individually under the present agrarian reform and whether a household holds other assets, such as farmland acquired under a previous land reform. These findings suggest that there is no direct causal link between certification-supported farming and the nature of farmers' livelihood diversification. However, both for a long time coexisted and interacted with each other through the cooperative system.

6.7 Compatibility of certification-supported farming with diversified activities

6.7.1 Necessity of compatibility

Many members understood that no other cash crop was more lucrative than sugarcane was and, at the same time, that small-scale sugarcane monoculture did not have a bright future. Their anxiety corresponded to an explanation given by the staff of ATC regarding marketing difficulties:

> The associations, including the study cooperative, aim to sell more sugarcane to ATC from their communal lands and, if possible, from members' individual land. However, ATC is confronted with two difficulties in expanding the export of their organic Fairtrade sugar. One is the limit of the capacity of ATC's milling factory. The other difficulty is the increasing competition with similar types of organic sugar from other major sugar-producing countries such as Brazil. ATC exports processed *muscovado* sugar to both Fairtrade and other ethical markets in the North.[15] Even in ethical markets, a small country such as the Philippines cannot compete with large producer countries that can control the international market and offer cheaper prices. Consequently, it is difficult for ATC to buy more produce from the existing associations at a higher level of prices and for each association to accept more members.

Therefore, the members of the cooperative continually attempted to diversify their livelihoods to prepare for the future while taking advantage of certification-supported sugarcane production. They were unlikely to stop producing sugarcane as long as they gained profits and other fringe benefits from the cooperative under the auspices of ATC and ATFI.[16] They needed both certification-supported farming and diversified activities.

6.7.2 Mechanisms for compatibility

How certification-supported farming was compatible with member households' diversification can be explained by two mechanisms. The first is inferred from members' perception of benefits from the cooperative. As Table 6.3 indicates, the members' perceived benefits of the cooperative correspond with the phase of livelihood diversification that the member households had reached. All these benefits derive from certification-supported sugarcane production. The dividend was the

Table 6.3 Benefits of the cooperative to STARFA member households in the different phases of livelihood diversification (Negros, the Philippines)

Livelihood diversification	Benefits
Phase 1: For survival	– Dividends from the cooperative – Wages for physical work on the communal land – Microcredit[a] – Priority in the allocation of additional work (e.g., *carabao* rearing) – Priority in scholarships for members' children[a]
Phase 2: For the transition from survival to accumulation	– Dividends from the cooperative – Wages for physical work on the communal land[b] – Microcredit – Production loans for individual land
Phase 3: For accumulation	– Dividends from the cooperative – Microcredit – Production loans for individual land (Confined to members who operate farmland individually)

Notes:
[a] Microcredit and scholarships for children were funded from the cooperative's revenue.
[b] In accordance with their shorter work period, Phase 2 members' annual income from wages tended to be smaller than that of the Phase 1 members.

only benefit common to all members because an equal amount was distributed to each member. Wages were limited to those who worked on the 20-hectare communal land. Production loans funded by ATFI were limited to members who operated farmland individually.[17] Microcredit was also open to all members. Access to production loans and microcredit was significant for the cooperative members who were not allowed to borrow from banks without having received a landownership certificate. All members who had used production loans and microcredit agreed that these schemes had meant that they no longer had to borrow from informal lenders at high interest rates.

Although members appreciated different benefits, the lives of the six member households in Phase 1 of livelihood diversification depended on the cooperative to the greatest extent. Whereas members with farmland intended for future individual ownership were obviously at an advantage, the members without such additional land (i.e., in Phase 1) did not appear discontent with the disparity. This may be because priority was given to the poorer member households in the allocation of work opportunities inside the cooperative as well as in the distribution of fringe benefits such as bonuses for workers and scholarships for members' children.[18] In other words, certification-supported farming helped such cooperative-dependent households to diversify their livelihoods for survival *inside* the cooperative.

The majority of members (except for those in Phase 1) shared the perception that the communal land no longer yields significant income. However, they retained their membership not only to gain landownership but also to use dividends, microcredit, and production loans for the diversification of their income sources. As Table 6.4 shows, the majority of members interviewed made use of microcredit from the cooperative's revenue for a variety of purposes, including not only consumption but also diversified economic activities *outside* the cooperative. "S*ari-sari* stores" and "individual farming" were typical of the livelihood portfolio falling into Phases 2 and 3 of livelihood diversification. It is reasonable to interpret that income from certification-supported farming and financial benefits from the cooperative helped some member households diversify their livelihood opportunities for the transition from survival to accumulation and for accumulation. Of the eight members who had never received microcredit – the two exceptional non-farm workers, one in Phase 1, and five in Phase 2 – two of the members in Phase 2 clearly expressed their fears for repayment. Further research is necessary to identify conditions for the successful use of microcredit for diversification.

In sum, although certification-supported farming contributed to different phases of livelihood diversification, it played a particularly crucial role for the survival of poorer member households. At the same time, the existence of such cooperative-dependent households (in Phase 1) implies that the opportunity provided through organic and Fair Trade initiatives alone does not necessarily lead to a pathway out of poverty for all beneficiaries of the initiatives. The existence

Table 6.4 STARFA members' major purposes for microcredit (Negros, the Philippines)

Purpose[a]	*No. of households*
Food and daily necessities	10
Medical bills	7
Education of children	7
Merchandise for *sari-sari* stores (convenience stores)	6
Poultry and livestock	3
Other non-farming activities	2
Individual farming	1
Did not answer	2
Borrower households in total	**38**
Never borrowed	8
Not eligible[b]	8
Total	**54**

Source: Field survey

Notes:
[a]Although some members stated more than two purposes, only one purpose was counted as each member's major purpose.
[b]According to the internal regulations, those who had not yet paid for initial capital shares were not allowed to use the microcredit scheme.

of members in Phases 2 and 3 implies that the two initiatives contribute to the improvement of the beneficiaries' economic status by offering a stable basis for diversification into higher-value activities, although whether they can start such activities depends on other factors.

The second mechanism for compatibility is collective farming on communal land, which was maintained in parallel with member households' individual activities, both farming and non-farming. As Table 6.5 shows, how each household was involved in collective farming depended on the phase of livelihood diversification it had reached. Households in Phase 1 (for survival) were most intensely involved in collective farming. For instance, although the three adult members of JB's family (introduced in Subsection 6.6.2 in this chapter) worked for the communal land throughout the year, the number of working days decreased in the agricultural lean season, especially in August. During that time they were likely to sell a pig they raised. Another household in Phase 1 was engaged in charcoal making only before the harvest season.

Typical households in Phase 2 divided their workload between farming on the communal land, farming on individual land, and non-farm activities in consideration of seasonal work demands for each livelihood. AC (see Subsection 6.6.2 in this chapter), as the only male earner of his family, worked for the communal

Table 6.5 Stylized division of labor at the households of STARFA members (Negros, the Philippines)

Livelihood diversification	Actors in on-farm activities		Actors in off-farm activities
	Communal	*Individual*	
Phase 1: For survival	Household heads and other family members	None	Household heads and other family members, especially in the agricultural lean season
Phase 2: For the transition from survival to accumulation (the case of households with individual farmland)	Household heads and employed workers*	Household heads and other family members	Other family members throughout the year
Phase 3: For accumulation	Employed workers*	Employed workers	Household heads and other family members throughout the year

Note: *The operation of the communal land depended *partially* on employed workers from the perspective of members in Phase 2 and *exclusively* on employed workers from the perspective of those in Phase 3.

land on weekdays throughout the year, in parallel looking after his individual farmland on weekends. AC's wife took charge of the *sari-sari* store throughout the year and helped AC on the individual land during the peak season. Because AC had recently leased additional land, he would need to allocate more time to his individual farming.

For households practicing livelihood portfolios classified as falling into Phase 3 (for accumulation), farming on both communal and individual plots was undertaken by members in Phases 1 and 2 and by non-member workers. In the household of TT (Subsection 6.6.2 in this chapter) all the family members were already involved in economic activities separate from the cooperative. With respect to his individual farming, TT, the eldest in his family, only planned the land use, hired agricultural workers, and supervised their daily performance. He did not see any difficulty in hiring the necessary numbers of workers and negotiating wage rates with them. TT enjoyed the basic benefits from the cooperative while also managing his transportation business (Table 6.3). It was even technically possible to maintain the communal farmland under certification solely with non-member workers who did not benefit from the certification program, although such an approach may conflict with the original purpose of the Fair Trade movement, namely helping small and marginal producers. The case of this cooperative suggests that collective farming can be compatible with members' diversification.

Although the existing literature suggests difficulty in continuing sustainable farming practices in parallel with diversification away from farming, such difficulty was also removed through the collective farming on the communal land. Corresponding to the fact that it is generally easier to keep a plantation or commercial farm organic than to ensure continued compliance by separate small farming households (Gomez Tovar et al., 2005), in this case cooperative members and employed workers universally adopted and maintained environmentally supportive practices to comply with organic and Fairtrade certifications. As long as the operation of the communal land is maintained, organic farming will continue on the communal land despite the increase in members' non-farm activities. In addition, 29 of the 30 members adopted organic practices on their individual land. This is a surprisingly high rate compared with many unsuccessful cases of small farmers who tend to discontinue organic practices (see Gomez Tovar et al., 2005; Gonzalez and Nigh, 2005). The cooperative and its members were successful in sustaining organic practices on both their communal and individual farmlands in parallel with their non-farm activities.

6.8 Concluding remarks

In discussing the relationships between agricultural certification and diversification, it is necessary first to understand how producers of a real certified cooperative diversify their livelihoods. In the case study of a cooperative consisting of land reform beneficiaries in the Philippines, the members over time reached different phases of livelihood diversification – for survival, for the transition from survival to accumulation, and for accumulation – whereas the nature of diversification depended primarily on factors other than organic and Fairtrade certifications. Although the existing literature implies the possible conflict of certification-supported farming with diversification away from farming, as this case study suggests, diversification does not necessarily occur away from farming. In this case diversification of the households of cooperative members progressed both away from farming (on the communal land) and into farming (on individual lands). The direction of diversification depends on the opportunities and assets each household can access. The poor long-term prospects for sugarcane production drove cooperative members to diversify their livelihoods in addition to certification-supported farming. In other words, agricultural certification allowed its beneficiary producers to prepare for anticipated threats or alternative opportunities rather than to continue to rely on a certified product. Two mechanisms were identified for the coexistence of both certification-supported farming and individual diversification. First, financial benefits from the cooperative were working as a coping strategy for survival-level diversification and as a stable basis for accumulation-level diversification. Second, member households' diversified activities were compatible with certification-supported farming through the communal land scheme. Households in different phases of diversification could participate in and benefit from the operation of the communal land in parallel with their individual economic activities. This communal scheme also enabled the cooperative to continue sustainable farming practices regardless of members' diversification. The compatibility of certification-supported farming with diversified activities, observed in this case, eventually supports the counterargument of Hayes (2008) and Smith (2009), which states that Fair Trade might contribute to economic diversification and structural change, rather than LeClair's (2002) argument that Fair Trade has a negative impact on diversification (Section 6.1).

This case study suggests two general conditions under which agricultural certification can be used successfully in parallel with other diversified activities. First, beneficiary producers need to have an obvious reason for continuing *both* certification-supported farming and other diversified activities. Second, there should be some mechanism for enabling certification-supported farming to be compatible with other economic activities. When these two conditions are met, certification-supported farming can contribute to beneficiaries' overall livelihood strategies. The operation of communal land with organic and Fairtrade certifications appears to be an effective form of assistance to land reform beneficiaries who share a strong desire to be landowners. With this common objective, they may diversify their livelihoods both into and away from farming and attempt to maximize their income-generating opportunities.

Notes

1 In 2010 the Philippines was the fourth-highest remittance-receiving country in the world, below India, China, and Mexico (ADB, 2012, p. 6).
2 Ellis (2000, p. 10) proposes the following definition of livelihood itself: "A livelihood comprises the assets (natural, physical, human, financial and social capital),

the activities, and the access to these (mediated by institutions and social relations) that together determine the living gained by the individual or household."

3 For example, in Bangladesh the increase in access to farmland for the land poor (owning up to 0.2 hectares) was found in contrast with the decrease of cultivation by the land rich (owning 1 hectare and more) (Hossain, 2004, p. 4054).

4 There were many reasons behind this shift, such as "the political contentiousness of land redistribution struggles" and the "uncertain and unpredictable outcomes" of agrarian reform projects (Borras and Franco, 2007, p. 24).

5 ATC was first created when a Japanese NGO attempted to link Japanese ethical consumers directly with displaced sugar plantation workers in the aftermath of the 1980 sugar crisis, and it gradually grew into the Alter Trade Group together with ATFI (Cabilo, 2009, pp. 142–143, 150).

6 Larger sugarcane farms tend to have higher yields per hectare (Billing, 2003, p. 87; World Bank, 2009, p. 152).

7 The cooperative was originally called Santa Rita Farm Workers Association. The abbreviation STARFA derives from this original name.

8 Purok is the lowest administrative unit under Barangay.

9 The certificate is officially called the Certificate of Land Ownership Award (CLOA).

10 According to information from ATFI, the cultivable land size per member was 0.27 hectares in this cooperative, whereas it ranged from 0.75 to 2.29 hectares in the other associations (as of August 2011).

11 Santa Rita is located in a flat lowland area about 10 kilometers inland from the capital city of the province, Bacolod. Although the village has comparatively good access to Bacolod, in the monsoon seasons (June to October) frequently flooded dirt roads make it difficult for villagers to move around inside the village.

12 *Carabaos* play an important role in traditional farming systems in the Philippines, being used "to plough, harrow, level land, puddle rice fields and thresh rice; pull carts, sledges, and logs; . . . and serve as riding or pack animals" (Momongan, Parker, de los Santos, and Ranjhan, 1989, p. 190).

13 An individual certificate is, in principle, limited to a maximum of 3.0 hectares per person under CARP (World Bank, 2009, p. 22).

14 ATC purchased sugarcane grown on the individual land of cooperative members only when the communal land was not able to provide the required quantities.

15 For example, in 2008 ATC sold 53 percent of its total production to the Fairtrade market, 36 percent to other overseas ethical markets, and 11 percent to the domestic market. A single price was applied when ATC purchased from the affiliated associations. The Philippines accounts for only 1 percent of world sugar production (FAO, 2010).

16 In 2012 the long partnership between ATFI and ATC came to an end. The 14 associations were unfortunately divided into 2 groups, with each group allied either to ATFI or to ATC. The nine associations that chose ATFI, including STARFA, automatically lost export opportunities. They kept their Fairtrade certification, but were selling their sugarcane to local conventional markets and were planning to process raw sugar by themselves for export to the Fairtrade market (as of February 2015) (Makita, 2016, pp. 197–198).

17 ATFI provided all affiliated associations with production loans. Although the other associations used such loans for their communal farmlands, in the cooperative under study loans were divided among members who wanted to use them for their individual farming according to internal decisions. This may reflect the decreased significance of communal farming in this cooperative.

18 The scholarship was funded from the cooperative's revenue.

7 Conclusion

7.1 Contradiction and tension: An analytical perspective

Our exploration of the convergence of Fair Trade and organic initiatives through these five case studies sheds light on various dimensions of the two initiatives as they have been applied within Asian agriculture. While each of these initiatives reflects a contradiction or tension between its two forms – as a movement and as certification – these case studies revealed that such contradictions can be more complex when the two initiatives are combined. What actually happens when they converge was revealed by attending to the different types of emergent contradictions or tensions.

At the outset, it should be noted that there is a fundamental difference between Fair Trade and organic certification systems. Whereas small farmers, when applying for a Fair trade certification, are required to choose a specific cash crop, when they apply for organic certification they are required to keep their entire production units organic. As the Kerala case study (Chapter 2) revealed, this difference in the unit of certification may generate a contradiction in real production settings. Among small farmers there are some who cannot simultaneously commit to the production of a selected cash crop while keeping their farmland organic. If they conduct organic farming they would wish to cultivate a variety of crops to gain a year-round income. However, if they have to focus on one Fair Trade crop they may need to increase the yield of the crop through the use of chemical inputs.

As Chapter 3 has shown, differences between the Fair Trade and organic initiatives relating to their core movements are more apparent in the context of GM seeds in the cotton sector than those revealed by a simple comparison of the two. Organic certification does not permit the use of GM seeds because the organic movement is aimed at conserving natural resources. Given its objective of helping disadvantaged producers, the Fair Trade movement cannot take such a definitive stand toward GM seeds as long as the majority of small farmers rely on these seeds. With GM seeds becoming increasingly widespread, the organic movement, which is primarily oriented toward environmental conservation, may not be compatible with the Fair Trade movement, which is primarily oriented toward economic development.

The process of obtaining double certification appears to be a smoother one for tea plantations than it is for small farmer groups. In the Darjeeling case study (Chapter 4), no tension was observed between the two initiatives. Instead, the ever-present tension that prevails between the plantation management and workers is the key factor that determines how workers benefit from certification. A contradiction emerges when Fairtrade certification is applied to the plantation sector. This relates to the fact that the management makes all decisions concerning the acquisition and maintenance of certification, while Fairtrade premiums – a benefit resulting from certification – are intended for workers. It is extremely difficult to overcome this contradiction under the power relations that have influenced every aspect of plantation work and life. However, Fairtrade certification will never result in substantial benefits for plantation workers unless this contradiction is resolved through some form of external intervention.

The Thai case study (Chapter 5) was presented as a unique case wherein the usual contradiction that exists between the movement and certification was not inevitable. Because Fair Trade and organic initiatives in Thailand originate from homegrown social movements seeking to create a sustainable society, the confluence of the two forms of certification is a natural one. Whereas business-oriented organic rice production has flourished in the region, some farmer groups have retained the original visions of their Buddhist-led movements.

In the case of the Philippines as well (Chapter 6), no contradiction was evident in the area of certification-supported agriculture, either between the Fair Trade and organic initiatives or between a movement and certification. This case study demonstrated a successful example of double certification. In this case tension was anticipated between the certification-supported farming and other diversified livelihood activities of small farmers. Although mechanisms existed for achieving compatibility between them, the findings of the case study also suggested that in the absence of such mechanisms the Fair Trade and organic initiatives may not help small farmers to enhance living.

7.2 Double certification in practice

Table 7.1 provides a summary of the five cases. Although these cases do not cover all of the plausible circumstances of double certification adoption, when reviewed from a cross-sectional perspective they enable us, to some extent, to address the three questions raised in Chapter 1, as discussed here.

7.2.1 Under what conditions can producers obtain double certification?

We observed both successful and unsuccessful cases of double certification. In two out of the five cases double certification was not realized. In the first case study conducted in Kerala, India, a key requirement of Fairtrade certification – namely, focusing on a specific cash crop – was not compatible with organic certification standards. In the second case study conducted in Telangana, India, the impact of

Table 7.1 Summary of the case studies

	Chapter 2	Chapter 3	Chapter 4	Chapter 5	Chapter 6
Place	Kerala, India	Telangana, India	Darjeeling, India	Northeast Thailand	Negros, the Philippines
Product	Coffee	Cotton	Tea	Rice	Sugar
Certified body	Cooperative	Cooperative	Plantation	Cooperative	Cooperative
Certification	Not certified; pursuing both	Fairtrade certified; pursuing organic certification	Double certified (organic first)	Double certified (organic first)	Double certified (organic first)
Contradictions studied	Between crop and production units	Between Fair Trade and organic movements	Between managers as the decision makers and workers as the beneficiaries	Between movements and certification	Between certification-supported farming and other livelihoods
External assistance for Fair Trade production	A series of activities to facilitate coffee production and export	None	None	None	None
External assistance for organic farming	Bio-liquid fertilizer	Non-Bt seeds; Fairtrade premiums	Fairtrade premiums	Fairtrade premiums; rice milling facilities	Fairtrade premiums; production loans
Influential context	A competing cash crop (rubber)	The spread of genetically modified seeds	Traditional power relations	Buddhist-based social movements	Land reform; agrarian change

Fairtrade certification, contrary to expectations, nullified the organic standard. In the remaining three cases organic certification was introduced first, and Fairtrade certification was subsequently obtained after the concerned small farmer groups and the plantation had fulfilled the requirement of organic certification. Parvathi and Waibel (2016, p. 210) also showed that in the case of double-certified black pepper farmers in India, organic certified farmers subsequently obtained additional Fairtrade certification.

These different outcomes suggested that small farmers benefit from Fair Trade–organic double certification only when they can afford to convert to organic farming. In other words, whereas it is comparatively easy for a small farmer group to obtain a Fair Trade certification if it has already been certified as organic, the reverse is fairly difficult. Given that organic certification is more accessible to larger farmers (e.g., Gomez Tovar et al., 2005), small farmers require special assistance when undergoing the process of conversion, such as access to production loans and full support extended by NGOs, as the Philippine and Thai case studies have revealed. Moreover, our case studies suggested that a Fair Trade certification does not always facilitate farmers' adopting organic farming.

7.2.2 Do small farmers prefer single or double certification?

There is no simple answer to the question of whether small farmers prefer single or double certification. Double certification is, in theory, preferable to single certification from the sales perspective. However, in reality double certification is not suitable for all farmers. Based on the premise that the majority of small farmers find it difficult to pursue organic certification, in addition to a Fair Trade certification, traders and consumers in the North are required to clearly differentiate the Fair Trade market category from other categories of ethical marketing. Even if Fair Trade-certified farmer groups use some chemical inputs and GM seeds for cultivating their products, such products should be regarded as sufficiently marketable as long as their sales to the Fair Trade market benefit the livelihoods of disadvantaged producers.

On the other hand, traders and consumers should be cognizant of the fact that the Fair Trade certification system, which is usually regarded as a tool for helping disadvantaged farmers, may not work for them in some contexts. In such contexts the organic certification system may be more appropriate and compatible with small farmers' livelihood strategies than is the Fair Trade certification system, as evidenced in the case of Kerala. If a Fair Trade certification is pursued in such situations, it is likely to result in the exclusion of its originally intended small farmers and instead benefit the wealthier farmers who are not its target group. Before introducing any certification program, practitioners should carefully examine which type of certification is most beneficial to the target group of the program.

7.2.3 How do the two forms of certification interact
with each other?

Our case studies revealed that the two forms of certification introduced to producers interact with each other in different ways. When organic certification is pursued first, producers' experiences in organic cultivation enable them to meet the environmental requirements of most Fair Trade certification programs. Activities for and under organic certification generally exert a positive influence on the acquisition and maintenance of a Fair Trade certification.

Conversely, the introduction of a Fair Trade certification program does not necessarily serve as an incentive to engage in organic farming. As the Kerala case revealed, the requirement to focus on a specific crop to obtain a Fair Trade certification may not be welcomed by small and marginal farmers who prefer to simultaneously cultivate multiple crops. This finding suggested a constraint related to the introduction of a Fair Trade certification program among small farmers. However, when this is successfully implemented among small farmers, the resulting social premiums (in the case of Fairtrade certification) can be used to promote organic farming (Ronchi, 2002, p. 17; Chapters 3, 4, 5, and 6 in this book). Although Fairtrade premiums were also used against organic farming in the case of Telangana, this resulted from the interaction of Fairtrade certification with GM seeds, and not with organic certification.

7.3 A variety of factors that influence the convergence of the two initiatives

The case studies also revealed the strong influence of particular contexts in which the initiatives are implemented on their confluence. Although such contexts are usually created by factors that are unique to specific countries or regions, and to the selected cash crops, our case studies revealed further influential factors relating to the two initiatives. These were (a) availability of competing cash crops, (b) agricultural innovation, (c) traditional power relations, (d) earlier social movements, and (e) ongoing agrarian change. While factors (a) and (b) directly influence income-generating opportunities for farmers, factors (c), (d), and (e) influence production practices indirectly, but significantly.

In the case of Kerala, the presence of rubber, another promising crop, was one of the reasons why small farmers were hesitant to focus on coffee for Fairtrade certification. It is highly likely that when another cash crop is available in a region, small farmers in this region will compare anticipated benefits from cultivating the Fairtrade crop with those obtained from cultivating the competing crop. In such cases small farmers, who require year-round incomes, are not always inclined to export one particular crop to ethical markets located in the North. Conversely, farmers are inclined to accept a Fair Trade certification program when they depend exclusively on one cash crop, such as cotton in the Telangana case, sugarcane in the Philippine case, and jasmine rice in the Thai case.

At present, the commercialized production of GM seeds constitutes the most significant innovation within the agricultural sector. As the case study of cotton cultivation in Telangana shows, whereas this innovation in agricultural production is a strong competitor to organic farming, it can be compatible with the Fair Trade initiative. With the spread of GM seeds, it no longer seems possible to pursue double certification. Policy makers and practitioners who intend to help small farmers through the provision of certification must choose between the following options: (a) tolerate GM seeds together with the introduction of a Fair Trade certification program, or (b) introduce organic certification together with financial support.

It seems to be much easier for plantations to obtain both forms of certification than for groups of individual producers to do so. However, whether plantation workers actually benefit from these forms of certification is another question. Given that the share of the plantation sector within Asia's Fairtrade-certified agriculture is significant (see Chapter 1), prevailing traditional plantation cultures should be carefully considered when introducing Fairtrade certification, as well as other interventions targeting workers, in plantations. Patron–client relationships established between plantation managers and workers are an influential component of these cultures. As the case of Darjeeling suggested, this patron–client relationship may hinder the egalitarian distribution of the benefits of Fairtrade certification to all plantation workers.

The Thai case study showed another example of traditional cultures. Long-established social movements in Thailand (partly supported by prevailing Buddhist values) contributed significantly to the coexistence of the two initiatives, even before the introduction of certification systems. While the principles of the Fair Trade and organic initiatives were developed in the West, they are perceived differently in Thailand. Through their participation in the two initiatives, some Thai producers and consumers may not only enhance their mutual business relationships, but may also attempt to jointly realize Buddhist teachings in their society.

The remaining factor is the marked and ongoing agrarian change in rural Asia, entailing an increasing shift from farming to non-farming economic activities. Such a shift may lead to the failure of many agricultural certification programs. However, despite the shift, small farmers have never abandoned farming in the context of land reforms. Instead, they attempt to diversify their livelihoods on the one hand, and to maximize profits from farming on the other. The case study of the Philippines revealed that the Fair Trade and organic initiatives may be conducive to such livelihood strategies pursued by participant farmers. Part-time small farmers can take advantage of Fair Trade and organic certification programs by making certification-supported farming compatible with their other livelihood activities. Evidently, agricultural certification interacts with several factors both within and beyond the agricultural sector.

7.4 Double certification and rural development in the South

In the final section of this chapter we will focus on a pending discussion topic, namely, the interpretation of Fair Trade–organic double certification from the perspective of rural development policies. Apart from being a marketing tool, double certification may have another purpose. Although the Fair Trade and organic initiatives have been harmonized with the existing movements in traditional Thai society, in general the two certification systems have been introduced to newly bring their respective associated movements into the rural South. This then raises the question of what the coexistence of the two movements means for rural development in the global South. This query further draws our attention to the fact that the pivotal question of how to *simultaneously* realize two crucial objectives of rural development – poverty reduction and environmental conservation – remains an open one (Sanderson, 2005). While the Fair Trade initiative is more oriented toward poverty reduction, the organic initiative is more oriented toward environmental conservation. Thus, the convergence of the two initiatives can, in theory, provide an answer to this open question. Our case studies revealed that a double certification program could possibly function as an intermediary institution, linking poverty reduction with environmental conservation.[1] However, in reality there are many challenges and constraints associated with achieving a convergence of the two movements on the side of producers.

A further question that is raised is how the two key objectives of rural development can be linked through double certification. The conceptual typology of the relationships between poverty reduction and biodiversity conservation, developed by Adams et al. (2010), appears to be useful in addressing this question. These authors present four different ways of looking at the connections and disconnections between poverty reduction and conservation. These standpoints are as follows: "(a) Poverty and conservation are separate policy realms; (b) poverty is a critical constraint on conservation; (c) conservation should not compromise poverty reduction; and (d) poverty reduction depends on living resource conservation" (Adams et al., 2010, p. 21).

The approach adopted by both the Fair Trade and organic initiatives clearly differs from that described in standpoint (a), which is usually adopted to ensure the maintenance of the populations of vulnerable species (Adams et al., 2010). Conservation is the primary goal in standpoints (b) and (c). Whereas poverty reduction is viewed "simply as a means to achieve more effective conservation" (Adams et al., 2010, p. 22) in standpoint (b), it is accorded even less importance in standpoint (c). As long as Fair Trade aims to offer "better trading conditions to, and securing the rights of, marginalized producers and workers" (WFTO and FLO, 2009, p. 4), (b) and (c) would not apply in the case of Fair Trade. In contrast to standpoint (b), standpoint (d) views "conservation strategies based on the sustainable use [of natural resources] primarily as a *means* [italics added] to reduce poverty" (Adams et al., 2010, p. 24). This standpoint most closely appears to match the objectives of Fair Trade–organic double certification as attempts to promote "the sustainable use of natural resources being a foundation of strategies for achieving poverty reduction and social justice" (Adams et al., 2010, p. 23). In other words, through the implementation of double certification environmental conservation can become a *means* of poverty reduction, which is the *end*. As the summary provided in Figure 7.1 indicates, double certification can be conceived as an institution linking environmental conservation and poverty reduction at the policy level, and sustainable production practices and enhanced incomes (and other benefits) at the producer level. Although single certification – either Fair

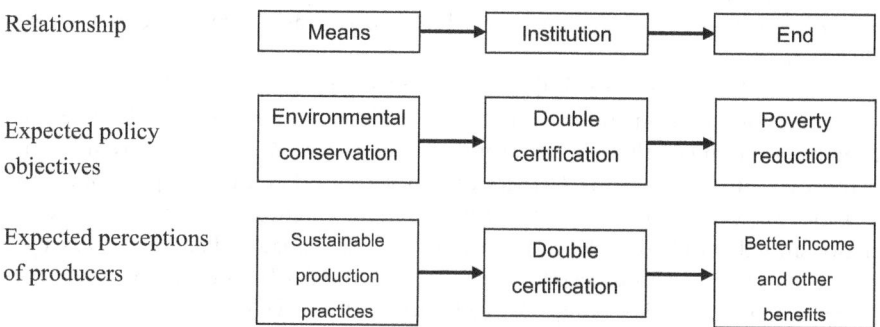

Figure 7.1 Fair Trade and organic initiatives in rural development
Source: Modified from Figure 1 in Makita (2016, p. 191)

Trade or organic – may also serve the same function (Makita, 2016), double certification links the two policy objectives more strongly than single certification does.

As Adams et al. (2010, p. 23) argued, environmental benefits are "a secondary gain" in standpoint (d). They further argued that "[c]onservation in response to this position tends toward the maintenance of . . . ecosystems rather than the preservation of biodiversity" (Adams et al., 2010, p. 23). Consequently, it is difficult to measure the impact of Fair Trade and organic initiatives on natural resources. This may be the reason underlying the conclusion reached by Blackman and Rivera (2011) that no strong correlation exists between Fair Trade certification and environmental benefits. As long as poverty reduction is the *end*, the impact of certification should be measured in terms of this end goal. The Fair Trade and organic initiatives can contribute to the conservation of natural resources only by changing producers' customary farming practices into ones that are more environmentally friendly.

If Fair Trade–organic double certification is successfully introduced to small farmers' groups, it is highly likely that small farmers will regard sustainable production practices as the *means* to improve their livelihoods. There are at least two conditions that need to be fulfilled for the means–end relationship to work through the implementation of double certification. First, it is critically important to define the end, that is, "poverty reduction at the producer level" in accordance with target beneficiaries' livelihood strategies. As suggested by the Kerala case study, which demonstrated an unsuccessful outcome, poverty reduction at the producer level does not necessarily imply an increase in total annual income. Some farmers may accord priority to securing a year-round income rather than to the amount of income itself that is obtained. In the case study of the Philippines, where the outcome was successful, the incentive to engage in sustainable production practices was not associated with income generated as a result of such production practices. Rather, it was associated with the compatibility of certification-supported farming with other income-generating activities. When the *end* is meaningful for small farmers, double certification will work as an intermediary institution.

The second condition for fulfilling the means–end relationship between environmental conservation and poverty reduction is that the means should be more attractive to farmers than other available means aimed at poverty reduction. In the case of Kerala, many farmers considered another cash crop, rubber, to be a more appealing means and did not, therefore, enter into the means–end relationship through double certification. In Telangana, India, which was the other example of an unsuccessful outcome, *unsustainable* production practices (conventional cultivation using Bt seeds) ironically served as a means to more effectively achieve better incomes than *sustainable* production practices (organic cultivation using non-Bt seeds). When strong competitors are present, the means intermediated by double certification will not lead to the desired end unless some value is added to the means.

We do not intend to assert that Fair Trade–organic double certification will contribute simultaneously to environmental conservation and poverty reduction under any circumstances. Rather, we argue that double certification suggests a potential means–end relationship that could function as a rural development framework to promote a win-win situation. Future studies should be conducted not only to clarify the conditions under which double certification works as the intermediary institution, but also to seek other schemes, devices, or innovations that can serve as intermediary institutions linking environmental conservation (the means) with poverty reduction (the end). For instance, the same means–end relationship may be observed in some non-certified Fair Trade schemes implemented between specific producers and buyers on a small scale. In the plantation sector, the role of a third-party body seems to be more crucial than double certification itself in order to make the means–end relationship work not for managers but for workers (see Chapter 4). Any such intermediary institution is expected to motivate small and marginal producers and workers to conserve natural resources on which their livelihoods depend for their own benefit.

Note

1 Following Leach, Mearns, and Scoones (1999, p. 226), *institutions* are defined in this chapter as "mediators of people–environment relations" and "regularized patterns of behavior between individuals and groups in society rather than . . . community-level organizations."

Bibliography

Aarat, S. (2013). *Senthaang chaawnaa khaakhaaw* [The path taken by rice-trading farmers]. Surin, Thailand: Rice Fund Surin Organic Agriculture Cooperative.

Adams, W. A., Aveling, R., Brockington, D., Dickson, B., Elliot, J., Hutton, J., et al. (2010). Biodiversity conservation and the eradication of poverty. In D. Roe and J. Elliott (Eds.), *The Earthscan reader in poverty and biodiversity conservation* (pp. 18–26). London: Routledge.

Alawattage, C., and Wickramasinghe, D. (2009). Institutionalisation of control and accounting for bonded labour in colonial plantations: A historical analysis. *Critical Perspectives on Accounting, 20*(6), 701–715.

Allen, J. C., and Malin, S. (2008). Green entrepreneurship: A method for managing natural resources? *Society and Natural Resources, 21*, 828–844.

Allen, P., and Kovach, M. (2000). The capitalist composition of organic: The potential of markets in fulfilling the promise of organic agriculture. *Agriculture and Human Values, 17*(3), 221–232.

Amekawa, Y. (2010). Rethinking sustainable agriculture in Thailand: A governance perspective. *Journal of Sustainable Agriculture, 34*(4), 389–416.

Ang, A. P., Sugiyarto, G., and Jha, S. (2009). *Remittances and household behavior in the Philippines.* Manila: ADB.

Asian Development Bank (ADB). (2005). *Poverty in the Philippines: Income, assets and access.* Manila: ADB.

ADB. (2009). *Poverty in the Philippines: Causes, constraints and opportunities.* Manila: ADB.

ADB. (2012). *Global crisis, remittances and poverty in Asia.* Manila: ADB.

Bacon, C. (2005). Confronting the coffee crisis: Can Fairtrade, organic, and specialty coffees reduce small-scale farmer vulnerability in Northern Nicaragua? *World Development, 33*(3), 497–511.

Bacon, C., Méndez, E., and Fox, J. (2008). Cultivating sustainable coffee: Persistent paradoxes. In C. Bacon, E. Méndez, S. Gliessman, D. Goodman, and J. Fox (Eds.), *Confronting the coffee crisis: Fair trade, sustainable livelihoods and ecosystems in Mexico and Central America* (pp. 337–372). Cambridge, MA: MIT Press.

Baffes, J. (2011). *Cotton, biotechnology, and economic development* (Policy Research Working Paper 5896). Washington, DC: The World Bank.

Barham, B. L., Callenes, M., Gitter, S., Lewis, J., and Weber, J. (2011). Fairtrade/organic coffee, rural livelihoods, and the "Agrarian Question": Southern Mexican coffee families in transition. *World Development, 39*(1), 134–145.

Barrett, H. R., Browne, A. W., Harris, P. J. C., and Cadoret, K. (2002). Organic certification and the UK market: Organic imports from developing countries. *Food Policy, 27*(4), 301–318.

Bassett, T. J. (2010). Slim pickings: Fairtrade cotton in West Africa. *Geoforum, 41*(1), 44–55.

Becchetti, L., Conzo, P., and Gianfreda, G. (2009). *Market access, organic farming and productivity: The determinants of creation of economic value on a sample of fair trade affiliated Thai farmers* (Econometica Working Papers 5). Milan: EconomEtica.

Behal, R. P. (2010). Coolie drivers or benevolent paternalists? British tea planters in Assam and the indenture labour system. *Modern Asian Studies, 44*(1), 29–51.

Behal, R. P., and Mohapatra, P. P. (1992). "Tea and money versus human life": The rise and fall of the indenture system in the Assam tea plantations 1840–1980. *Journal of Peasant Studies, 19*(3 and 4), 142–172.

Bennett, M., and Franzel, S. (2013). Can organic and resource-conserving agriculture improve livelihoods? A synthesis. *International Journal of Agricultural Sustainability, 11*(3), 193–215.

Besky, S. (2014). *The Darjeeling distinction: Labor and justice on fair-trade tea plantations in India*. Berkeley and Los Angeles, CA: University of California Press.

Best, H. (2008). Organic agriculture and conventionalization hypothesis: A case study from West Germany. *Agriculture and Human Values, 25*(1), 95–106.

Bhaumik, S. K. (2007). Growth and composition of rural non-farm employment in India in the era of economic reforms. *Indian Economic Journal, 55*(3), 40–65.

Billing, M. S. (1993). "Syrup in the wheels of progress": The inefficient organisation of the Philippine sugar industry. *Journal of Southeast Asian Studies, 24*(1), 122–147.

Billing, M. S. (2003). *Barons, brokers, and buyers: The institutions and cultures of Philippine sugar*. Honolulu, HI: University of Hawai'i Press.

Blackman, A., and Rivera, J. (2011). Producer-level benefits of sustainability certification. *Conservation Biology, 25*(6), 1176–1185.

Bolwig, S., Gibbon, P., and Jones, E. S. (2009). The economics of smallholder organic contract farming in Tropical Africa. *World Development, 37*(6), 1094–1104.

Borras, S. M. (2001). State – society relations in land reform implementation in the Philippines. *Development and Change, 32*, 531–561.

Borras, S. M., Carranza, D., and Franco, J. C. (2007). Anti-poverty and anti-poor? The World Bank's market-led agrarian reform experiment in the Philippines. *Third World Quarterly, 28*(8), 1557–1576.

Borras, S. M., and Franco, J. C. (2007). *The national campaign for land reform in the Philippines*. Brighton, UK: Institute of Development Studies.

Bowes, J., and Croft, D. (2007). Organic and Fair trade crossover and convergence. In S. Wright and D. McCrea (Eds.), *The handbook of organic and fair trade food marketing* (pp. 262–283). Oxford, UK: Blackwell Publishing.

Breman, J. (2000). Labour and landlessness in South and South-East Asia. In D. Bryceson, C. Kay and J. Mooij (Eds.), *Disappearing peasantries?* (pp. 231–246). London: ITDG Publishing.

Browne, A. W., Harris, P. J. C., Hofny-Collins, A. H., Pasiecznik, N., and Wallace, R. R. (2000). Organic production and ethical trade: Definition, practice and links. *Food Policy, 25*(1), 69–89.

Buchenrieder, G. (2005). Non-farm rural employment – Review of issues, evidence and policies. *Quarterly Journal of International Agriculture, 44*(1), 3–18.

Buck, D., Getz, C., and Guthman, J. (1997). From farm to table: The organic vegetable commodity chain of northern California. *Sociologia Ruralis, 37*(1), 3–20.

Buddhadasa, B. (1986). *Dhammic socialism.* Bangkok: Thai Inter-Religious Commission for Development (TICD).

Cabilo, Z. M. D. (2009). From North to South: Campaigning for fairtrade in the Philippines. In T. S. E. Tadem (Ed.), *Localizing and transnationalizing contentious politics: Global civil society movements in the Philippines* (pp. 137–180). Lanham, MD: Rowman and Littlefield Publishers.

Carney, C. P. (1989). International patron–client relationships: A conceptual framework. *Studies in Comparative International Development, 24*(2), 42–55.

Carpenter, D. (2003). An investigation into the transition from technological to ecological rice farming among resource poor farmers from the Philippine island of Bohol. *Agriculture and Human Values, 20*(2), 165–176.

Chakrabarti, A. M., and Sarkar, K. (2005). *Productivity and labour welfare: A study in tea gardens.* Kolkata, India: State Labour Institute, Government of West Bengal.

Chatterjee, P. (2001). *A time for tea: Woman, labor and post/colonial politics on an Indian plantation.* Chapel Hill, NC: Duke University Press.

Choudhary, B., and Gaur, K. (2010). *Bt cotton in India: A country profile.* New Delhi, India: The International Service for the Acquisition of Agri-biotech Applications (ISAAA).

Choudhary, B., and Gaur, K. (2011). *Adoption and impact of Bt cotton in India, 2002 to 2010.* New Delhi, India: ISAAA.

Coombes, B., and Campbell, H. (1998). Dependent re-production of alternative modes of agriculture: Organic farming in New Zealand. *Sociologia Ruralis, 38*(2), 127–145.

Cottingham, M., and Winkler, E. (2007). The organic consumer. In S. Wright and D. McCrea (Eds.), *The handbook of organic and fair trade food marketing* (pp. 29–53). Oxford, UK: Blackwell Publishing.

Dankers, C. (2003). *Environmental and social standards, certification and labeling for cash crops.* Rome: FAO.

Dasgupta, R. (1992). Plantation labour in colonial India. *Journal of Peasant Studies, 19*(3 and 4), 173–196.

De Wit, J., and Verhoog, H. (2007). Organic values and the conventionalization of organic agriculture. *NJAS – Wageningen Journal of Life Sciences, 54*(4), 449–462.

Dhavse, R. (2004). *Organic: An easier choice.* Retrieved March 15, 2009, from www.indiatogether.org/2004/apr/agr-biofert.html

Diprose, G., and McGregor, A. (2009). Dissolving the sugar fields: Land reform and resistance identities in the Philippines. *Singapore Journal of Tropical Geography, 30,* 52–69.

Dowd, B. M. (2008). Organic cotton in Sub-Saharan Africa: A new development paradigm? In W. G. Moseley and L. C. Gray (Eds.), *Hanging by a thread: Cotton, globalization, and poverty in Africa* (pp. 251–271). Athens, OH: Ohio University Press.

Dowd-Uribe, B. M., and Bingen, J. (2008). Debating the merits of biotech crop adoption in sub-Saharan Africa: Distributional impacts, climatic variability and pest dynamics. *Progress in Development Studies, 11*(1), 63–68.

East Asia Rice Working Group. (2006). *Rice farming: Best practices from selected Asian countries.* Quezon City, Philippines: Rice Watch and Action Network.

Edwards, N. (2013). Values and the institutionalization of Indonesia's organic agriculture movement. In F. Michele (Ed.), *Social activism in Southeast Asia* (pp. 72–88). Oxon, UK: Routledge.

Eernstman, N., and Wals, A. E. J. (2009). Interfacing knowledge systems: Introducing certified organic agriculture in a tribal society. *NJAS – Wageningen Journal of Life Sciences, 56*(4), 375–390.

Ellis, F. (1998). Household strategies and rural livelihood diversification. *Journal of Development Studies, 35*(1), 1–38.

Ellis, F. (2000). *Rural livelihoods and diversity in developing countries.* Oxford, UK: Oxford University Press.

Eyhorn, F. (2007). *Organic farming for sustainable livelihoods in developing countries? The case of cotton in India.* Zurich, Switzerland: Vdf Hochschulverlag AG an der ETH.

Fairtrade Foundation. (2011). *Q and A: Fairtrade standards and genetically modified organisms (GM).* Retrieved June 14, 2011, from http://www.fairtrade.org.uk/includes/documents/cm_docs/2011/f/fairtrade_gm_q_a_jan_08.pdf

Fairtrade India. (2013). *Media introduction – Fairtrade India.* Retrieved November 3, 2014, from http://fairtradeindia.org/wp-content/uploads/2013/12/2013–11–21_Press-Release_Media-Introduction-Fairtrade-India.pdf

Fairtrade International (FLO). (2009). *Why fairtrade is unique.* Retrieved August 25, 2009, from http://www.fairtrade.net/why_fairtrade_is_unique.html

FLO. (2011a). *Benefits of fairtrade for producers.* Retrieved December 1, 2011, from http://www.fairtrade.net/cotton.html#c3795

FLO. (2011b). *Generic Fairtrade standard for small producer organizations.* Bonn, Germany: FLO.

FLO. (2011c). *Fairtrade standard for small producer organisations.* Bonn, Germany: FLO.

FLO. (2013). *Monitoring the scope and benefits of Fairtrade,* 5th ed. Bonn, Germany: FLO.

FLO. (2014a). *Monitoring the scope and benefits of Fairtrade,* 6th ed. Bonn, Germany: FLO.

FLO. (2014b). *Benefits of fairtrade.* Retrieved November 3, 2014, from http://www.fairtrade.net/benefits-of-fairtrade.html

FLO. (2015). *Fairtrade standard for hired labor.* Retrieved November 6, 2015, from http://www.fairtrade.net/fileadmin/user_upload/content/2009/standards/documents/generic-standards/2015–08–01_EN_HL.pdf

FLO-CERT. (2015). *Fee system small producer organization (explanatory document).* Retrieved November 6, 2015, from http://www.flocert.net/wp-content/uploads/2014/03/PC-FeeSysSPO-ED-26-en1.pdf

FLO-CERT. (2016). *A list of FLO-certified groups in Thailand.* Retrieved February 25, 2016, from http://www.flocert.net/

Food and Agriculture Organisation of the United Nations (FAO). (2010). *FAO statistical yearbook 2010.* Rome: FAO.

FAO. (2014). *What is organic agriculture?* Retrieved November 3, 2014, from http://www.fao.org/organicag/oa-faq/oa-faq1/en/

Franco, J. C. (2008). Making land rights accessible: Social movements and political-legal innovation in the rural Philippines. *Journal of Development Studies, 44*(7), 991–1022.

French Fair Trade Platform. (2015). *International guide to fair trade labels.* Bondues, France: French Fair Trade Platform/Plate-Forme pour le Commerce équitable.

Frundt, H. J. (2009). *Fair bananas! Farmers, workers, and consumers strive to change an industry.* Tucson, AZ: University of Arizona Press.

Getz, C., and Shreck, A. (2006). What organic and fair trade labels do not tell us: Towards a place-based understanding of certification. *International Journal of Consumer Studies, 30*(5), 490–501.

Gibbs, D. (2000). Globalization, the bioscience industry and local environmental responses. *Global Environmental Change, 10*(4), 245–257.

Giovannucci, D., and Ponte, S. (2005). Standards as a new form of social contract? Sustainability initiatives in the coffee industry. *Food Policy, 30*(3), 284–301.

Glover, D. (2010). Exploring the resilience of Bt cotton's pro-poor success story. *Development and Change, 41*(6), 955–981.

GMO Compass. (2011). *GM crop production: GMO cultivation area by crop.* Retrieved December 12, 2011, from http://www.gmo-compass.org/eng/grocery_shop ping/crops/161.genetically_modified_cotton.html

Gohlert, E. W. (1991). *Power and culture: The struggle against poverty in Thailand.* Bangkok: White Lotus.

Goldstein, M. C., Childs, G., and Wangdui, P. (2008). "Going for income" in village Tibet: A longitudinal analysis of change and adaptation, 1997–2007. *Asian Survey, 48*(3), 514–534.

Gomez Tovar, L., Martin, L., Gomez Cruz, M. A., and Mutersbaugh, T. (2005). Certified organic agriculture in Mexico: Market connections and certification practices in large and small producers. *Journal of Rural Studies, 21,* 461–474.

Gonzalez, A. A., and Nigh, R. (2005). Smallholder participation and certification of organic farm products in Mexico. *Journal of Rural Studies, 21,* 449–460.

Gouse, M., Shankar, B., and Thirtle, C. (2008). The decline of Bt cotton in KwaZulu-Natal: Technology and institutions. In W. G. Moseley and L. C. Gray (Eds.), *Hanging by a thread: Cotton, globalization, and poverty in Africa* (pp. 103–120). Athens, OH: Ohio University Press.

Green Net. (2016). *About Green Net cooperative.* Retrieved May 7, 2016, from http://www.greennet.or.th/en/about/greennet

Guptill, A. (2009). Exploring the conventionalization of organic dairy: Trends and counter-trends in upstate New York. *Agriculture and Human Values, 26*(1–2), 29–42.

Guthman, J. (1998). Regulating meaning, appropriating nature: The codification of California organic agriculture. *Antipode, 30*(2), 135–154.

Guthman, J. (2002). Commodified meanings, meaningful commodities: Re-thinking production – consumption links through the organic system of provision. *Sociologia Rulais, 42*(4), 295–311.

Guthman, J. (2004a). *Agrarian dreams: The paradox of organic farming in California.* Berkeley, CA: University of California Press.

Guthman, J. (2004b). The trouble with "organic lite" in California: A rejoinder to the "conventionalization" debate. *Sociologia Ruralis, 44*(3), 301–316.

Haggblade, S., Hazell, P., and Readon, T. (2002). *Strategies for stimulating poverty-alleviating growth in the rural nonfarm economy in developing countries* (EPDT discussion paper no. 92). Washington, DC: International Food Policy Research Institute and the World Bank.

Haggblade, S., Hazell, P., and Readon, T. (2010). The rural non-farm economy: Prospects for growth and poverty reduction. *World Development, 38*(10), 1429–1441.

Halberg, N., and Muller, A. (2013). Organic agriculture, livelihoods and development. In N. Halberg and A. Muller (Eds.), *Organic agriculture for sustainable livelihoods* (pp. 1–20). Oxon, UK: Routledge.

Hall, A. (1974). Patron–client relations. *Journal of Peasant Studies, 1*(4), 506–509.

Hall, A., and Mogyrody, V. (2001). Organic farmers in Ontario: An examination of the conventionalization argument. *Sociologia Ruralis, 41*(4), 399–422.

Hart, G. (1994). The dynamics of diversification in an Asian Rice Region. In B. Koppel, J. Hawkins, and W. James (Eds.), *Development or deterioration? Work in rural Asia* (pp. 47–71). Boulder, CO: L. Rienner Publishers.

Hayes, M. G. (2008). "Fighting the tide: Alternative trade organizations in the era of global free trade" – A comment. *World Development, 36*(12), 2953–2961.

Herring, R. J. (2005). Miracle seeds, suicide seeds, and the poor: GMOs, NGOs, farmers and the state. In R. Ray and M. F. Katzenstein (Eds.), *Social movements in India* (pp. 203–232). New York and Toronto, Canada: Rowman & Littlefield Publishers.

Hisas, L., and Penaflor, P. (2006). *Implementation barriers to sustainable development: A civil society assessment in 15 countries in Asia, Africa and Latin America.* Kampala, Uganda: Sustainability Watch International Network.

Hossain, M. (2004). Rural non-farm economy: Evidence from household surveys. *Economic and Political Weekly, 39*(36), 4053–4058.

INDOCERT. (2011). *FAQ: Agriculture.* Retrieved December 10, 2011, from http://www.indocert.org/faq/agriculture.html

International Cotton Advisory Committee (ICAC). (2011). *Descriptions of production programs: Organic, fair trade, cotton made in Africa, and the better cotton initiative* (Attachment III to SC-N-509). Washington, DC: ICAC.

International Federation of Organic Agriculture Movements (IFOAM). (2007). *The IFOAM basic standards for organic production and processing (version 2005).* Bonn, Germany: IFOAM.

IFOAM. (2011). *Definition of organic agriculture.* Retrieved December 10, 2011, from http://ifoam.org/growing_organic/definitions/doa/index.html

IFOAM. (2014). *Principles of organic agriculture.* Retrieved November 3, 2014, from http://www.ifoam.org/en/principles-organic-agriculture/principle-fairness

International Fund for Agricultural Development (IFAD). (2005). *Organic agriculture and poverty reduction in Asia: China and India focus.* Rome: IFAD.

Ishii, Y. (1991). *Tai bukkyô nyûmon* [An introduction to Thai Buddhism]. Tokyo: Mekon.

Ito, T. (2012). *Modern Thai Buddhism and Buddhadasa Bhikkhu: A social history.* Singapore: NUS Press.

Izumi, O. (1995). Tai sonraku shakai ni okeru bukkyo no ichi doko: Tohoku tai no futari no kaihatsu so wo megutte [A trend of Buddhism in Thai rural society: Two development monks in Northeast Thailand]. *Gengo Chiiki Bunka Kenkyu, 1,* 81–104.

Jaffee, D. (2007). *Brewing justice: Fairtrade coffee, sustainability, and survival.* Berkeley, CA: University of California Press.

Jaffee, D., and Howard, P. H. (2010). Corporate cooptation of organic and fair trade standards. *Agriculture and Human Values, 27*(4), 387–399.

Jannuzi, F. T., and Peach, J. T. (1980). *The agrarian structure of Bangladesh: An impediment to development*. Boulder, CO: Westview Press.

Kaltoft, P. (1999). Values about nature in organic farming practice and knowledge. *Sociologia Ruralis, 39*(1), 39–53.

Kasem, S., and Thapa, G. B. (2012). Sustainable development policies and achievements in the context of the agriculture sector in Thailand. *Sustainable Development, 20*(2), 98–114.

Kaufman, A. H. (2012). Organic farmers' connectedness with nature: Exploring Thailand's alternative agriculture network. *Worldviews, Global Religions, Culture and Ecology, 16*(2), 154–178.

Kaufman, A. H., and Mock, J. (2014). Cultivating greater well-being: The benefits Thai organic farmers experience from adopting Buddhist eco-spirituality. *Journal of Agricultural Environmental Ethics, 27*(6), 871–893.

Kjosavik, D. J., and Shanmugaratnam, N. (2007). Property rights dynamics and indigenous communities in highland Kerala, South India: An institutional-historical perspective. *Modern Asian Studies, 41*(6), 1183–1260.

Knight, K. W., and Newman, S. (2013). Organic agriculture as environmental reform: A cross-national investigation. *Society and Natural Resources, 26*(4), 369–385.

Kolk, A. (2013). Mainstreaming sustainable coffee. *Sustainable Development, 21*(5), 324–337.

Krasachat, W. (2012). Organic production practices and technical inefficiency of Durian farms in Thailand. *Procedia Economics and Finance, 3*, 445–450.

Kuiper, M., Meijerink, G., and Eaton, D. (2008). Rural livelihoods: Interplay between farm activities, non-farm activities and the resource base. In R. P. Roetter, H. van Keulen, M. Kuiper, J. Verhagen, and H. H. van Laar (Eds.), *Science for agriculture and rural development in low-income countries* (pp. 77–95). Dordrecht, Netherlands: Springer.

La Trobe, H. L., and Acott, T. G. (2000). Localising the global food system. *International Journal of Sustainable Development and World Ecology, 7*(4), 309–320.

Lalitha, N., Bharat, R., and Viswanathan, P. K. (2009). India's experience with Bt cotton: Case studies from Gujarat and Maharashtra. In R. Tripp (Ed.), *Biotechnology and agricultural development: Transgenic cotton, rural institutions and resource-poor farmers* (pp. 135–167). Oxon, UK: Routledge.

Le Minoux, T. (2012). *Fair trade in Thailand*. Bangkok: Green Net Cooperative.

Leach, M., Mearns, R., and Scoones, I. (1999). Environmental entitlements: Dynamics and institutions in community-based natural resource management. *World Development, 27*(2), 225–247.

LeClair, M. S. (2002). Fighting the tide: Alternative trade organizations in the era of global free trade. *World Development, 30*(6), 949–958.

Lemarchand, R., and Legg, K. (1972). Political clientelism and development: A preliminary analysis. *Comparative Politics, 4*, 149–178.

Lewis, J., and Runsten, D. (2008). Is fairtrade – organic coffee sustainable in the face of migration? Evidence from a Oaxacan community. *Globalisations, 5*(2), 275–290.

Liamzon, C. M. (1996). Agrarian reform: A continuing imperative or an anachronism. *Development in Practice, 6*(4), 315–323.

Lindberg, J. (2012). The diversity and spatiality of rural livelihoods in Southern Sri Lanka: Access, poverty, and local perceptions. *Norwegian Journal of Geography, 66*(2), 63–75.

Lockie, S., and Halpin, D. (2005). The "conventionalization" thesis reconsidered: Structural and ideological transformation of Australian organic agriculture. *Sociologia Ruralis, 45*(4), 284–307.

Lumley, S. (2002). *Sustainability and degradation in less developed countries: Immolating the future?* Aldershot, UK: Ashgate.

Lyon, S., and Moberg, M. (Eds.). (2010). *Fair trade and social justice: Global ethnographies.* New York, NY: New York University Press.

Makita, R. (2007). *Livelihood diversification and landlessness in rural Bangladesh.* Dhaka, Bangladesh: University Press Limited.

Makita, R. (2016). A role of fair trade certification for environmental sustainability. *Journal of Agricultural and Environmental Ethics, 29*(2), 185–201.

Manas, L., and Prasit, K. (2007). The possibility of Hom Mali rice production in organic farming systems as an alternative farming career with poverty alleviation potential for lower-northeastern farmers: Surin Province case. *Khon Kaen Agriculture Journal, 35*(2), 177–188.

McCoy, A. W. (1992). Sugar barons: Formation of a native planter class in the colonial Philippines. *Journal of Peasant Studies, 19*(3/4), 106–141.

McNeely, J. A., and Scherr, S. J. (2003). *Ecoagriculture: Strategies to feed the world and save wild biodiversity.* Washington, DC: Island Press.

Missingham, B. D. (2003). *The assembly of the poor in Thailand: From local struggles to national protest movement.* Chiang Mai, Thailand: Silkworm Books.

Momongan, V. G., Parker, B. A., de los Santos, E. B., and Ranjhan, S. K. (1989). Breeding programs for improved draught animal power: Crossbreeding of Buffaloes. In D. Hoffmann, J. Nari, and R. J. Petheram (Eds.), *Draught animals in rural development* (pp. 190–194). Canberra, Australia: Australian Centre for International Agricultural Research.

Moral Rice. (2014). *Khwaam pen maa* [An outline history (of Moral Rice)]. Retrieved November 30, 2014, from http://www.moralrice.net/

Morera, M. C., and Gladwin, C. H. (2006). Does off-farm work discourage soil conservation? Incentives and disincentives throughout two Honduran hillside communities. *Human Ecology, 34*(3), 355–378.

Morse, S., Bennett, R., and Ismael, Y. (2007). Isolating the "farmer" effect as a component of the advantage of growing genetically modified varieties in developing countries: A Bt cotton case study from Jalgaon, India. *Journal of Agricultural Science, 145*(5), 491–500.

Muradian, R., and Pelupessy, W. (2005). Governing the coffee chain: The role of voluntary regulatory systems. *World Development, 33*(12), 2029–2044.

Mutersbaugh, T. (2002). The number is the beast: A political economy of organic-coffee certification and producer unionism. *Environment and Planning A, 34*(7), 1165–1184.

Mutersbaugh, T. (2005). Fighting standards with standards: Harmonization, rents and social accountability in certified agrofood networks. *Environment and Planning A, 37*(11), 2033–2051.

Mutersbaugh, T., and Klooster, D. (2011). Environmental certification: Standardization for diversity. In S. Lockie and D. Carpenter (Eds.), *Agriculture, biodiversity and markets: Livelihoods and agroecology in comparative perspective* (pp. 155–174). Oxon, UK: Routledge.

Narayanamoorthy, A., and Kalamkar, S. S. (2006). Is Bt cotton cultivation economically viable for Indian farmers? An empirical analysis. *Economic and Political Weekly, 41*(26), 2716–2724.

Natedao, T. (2011a). Contesting meanings in organic agriculture and the shifting identities of organic growers in Thailand. In V. Chayan and W. Chusak (Eds.), *Revisiting agrarian transformations in the Greater Mekong sub-region: New challenges* (pp. 89–115). Chiang Mai, Thailand: Regional Center for Social Science and Sustainable Development, Chiang Mai University.

Natedao, T. (2011b). The transition from conventional to organic rice production in Northeast Thailand: Prospect and challenges. In M. A. Stewart and P. A. Coclanis (Eds.), *Environmental change and agricultural sustainability in the Mekong Delta* (pp. 411–435). Dordrecht, The Netherlands: Springer.

Nature Care Society (NCS). (2012). *Eekasaan prakoop kaan prachum yai saaman pracam pii 2554* [Annual report, 2011]. Yasothon, Thailand: NCS.

Nayak, A. K., Gangwar, B., Shukla, A. K., Mazumdar, S. P., Kumar, A., Raja, R., et al. (2012). Long-term effect of different integrated nutrient management on soil organic carbon and its fractions and sustainability of rice – wheat system in Indo Gangetic Plains of India. *Field Crops Research, 127*, 129–139.

Neilson, J., Arifin, B., Gracy, C. P., Kham, T. N., Pritchard, B., and Soutar, L. (2011). Challenges of global environmental governance by non-state actors in the coffee industry: Insights from India, Indonesia and Vietnam. In S. Lockie and D. Carpenter (Eds.), *Agriculture, biodiversity and markets: Livelihoods and agroecology in comparative perspective* (pp. 175–200). Oxon, UK: Routledge.

Neilson, J., and Pritchard, B. (2009). *Value chain struggles: Institutions and governance in the plantation districts of South India.* Oxford, UK: Wiley-Blackwell.

Nelson, E., Gomez Tovar, L., Rindermann, R. S., and Gomez Cruz, M. A. (2010). Participatory organic certification in Mexico: An alternative approach to maintaining the integrity of the organic label. *Agriculture and Human Values, 27*(2), 227–237.

Newell, P. (2009). Bio-hegemony: The political economy of agricultural biotechnology in Argentina. *Journal of Latin American Studies, 41*(1), 27–57.

Nicholls, A., and Opal, C. (2005). *Fair trade: Market-driven ethical consumption.* London: Sage.

Nigh, R. (1997). Organic agriculture and globalization: A Maya associative corporation in Chiapas, Mexico. *Human Organization, 56*(4), 427–436.

Nishikawa, J., and Noda, M. (Eds.). (2001). *Bukkyo, kaihatsu, NGO: Tai kaihastusô ni manabu kyosei no chie* [Buddhism, development, and NGO: Wisdom of coexistence, lessons from development monks in Thailand]. Tokyo: Shinhyoron.

Nitasmai, T. (1996). Sustainable agriculture in Thailand. In P. Hirsch (Ed.), *Seeing forests for trees: Environment and environmentalism in Thailand* (pp. 268–286). Chang Mai, Thailand: Silkworm Books.

Nozaki, A. (1995). Tai no atarashii noson kaihatsu undo: Tohoku tai no kaihatsuso no jirei kenkyu [New movement for rural development in Thailand: A case study of development monks in Northeast Thailand]. *Tohokugakuin Daigaku Ronshu, Keizaigaku, 129*, 107–149.

Nuntana, U. (2001). *Fair trade in organic rice: A case study from Thailand.* Paper presented at the Annual Conference of Development Studies Association, University of Manchester, 10–12 September 2001.

Nuntana, U., and Winnett, A. (2002). Fair trade in organic rice: A case study from Thailand. *Small Enterprise Development, 13*(3), 45–53.

Nuntiya, H. (1998). *Thun chumchon: Botrian lae prasopkarn roongsiikhaaw chomrom rak thammachart* [Community capital: Lessons and experience from Nature Care Society rice mill]. Bangkok: Local Development Institute.

Nuntiya, H., and Thunya, S. (2011). Essential fundamentals and problems in the development of a green community: A case study of intimate trust in organic farming in Surin, Thailand. *GCOE International Joint Research, 6*, 48–80.

Office of Agricultural Economics (OAE). (2014). *Thailand foreign agricultural trade statistics*. Bangkok: Ministry of Agriculture and Co-operatives.

Orboi, M-D. (2013). Aspects regarding the evolution of the organic food market in the world. *Research Journal of Agricultural Science, 45*(2), 201–209.

Orjavik, K. (2012). World of organic certification 2012. In H. Willer and L. Kilcher (Eds.), *The world of organic agriculture: Statistics and emerging trends 2012* (pp. 137–141). Bonn: FiBL and IFOAM.

Osakwe, E. (2009). *Cotton fact sheet India*. Retrieved June 23, 2011, from http://www.icac.org/econ_stats/country_facts/e_india.pdf

Parnwell, M. (2005). The power to change: Rebuilding sustainable livelihoods in North-East Thailand. *The Journal of Transdisciplinary Environmental Studies, 4*(2), 1–21.

Parnwell, M., and Seeger, M. (2008). The relocalization of Buddhism in Thailand. *Journal of Buddhist Ethics, 15*, 79–176.

Parvathi, P., and Waibel, H. (2016). Organic agriculture and fair trade: A happy marriage? A case study of certified smallholder black pepper farmers in India. *World Development, 77*(1), 206–220.

Patil, R. R. (2002). An investigative report on circumstances leading to death among Indian cotton farmers. *International Journal of Occupational Medicine and Environmental Health, 15*(4), 405–407.

Pendergrast, M. (2015). *Beyond fair trade: How one small coffee company helped transform a hillside village in Thailand*. Vancouver, Canada: Greystone Books.

Phittaya, W. (1993). *Mura no shu niwa kari ga aru: Hotoku no kaihatsuso* [I am indebted to villagers: A development monk repaying villagers' kindness]. Nontaburi, Thailand: Sangsan Publishing.

Pinit, L. (2012). *Development monks in Northeast Thailand*. Kyoto: Kyoto University Press.

Pratt, J. (2009). Incorporation and resistance: Analytical issues in the conventionalization debate and alternative food chains. *Journal of Agrarian Change, 9*(2), 155–174.

Prawase, W. (1988). Buddhist agriculture and the tranquility of Thai society. In P. Seri and R. Bennoun (Eds.), *Turning point of Thai farmers* (pp. 1–43). Bangkok: Thai Institute for Rural Development.

Pretty, J. N., Morison, J. I. L., and Hine, R. E. (2003). Reducing food poverty by increasing agricultural sustainability in developing countries. *Agriculture, Ecosystems and Environment, 95*, 217–234.

The Provincial Government of Negros Occidental. (2007). *An evaluation of the 1988 comprehensive agrarian reform programme (CARP) implementation in Negros Occidental*. Negros Occidental, the Philippines: The Provincial Government of Negros Occidental.

Pye-Smith, C. (1997). *The Philippines: In search of justice*. Oxford, UK: Oxfam.

Qaim, M. (2010). Benefits of genetically modified crops for the poor: Household income, nutrition, and health. *New Technology, 27*(5), 552–557.

Ramasundaram, P., Vennila, S., and Ingle, R. K. (2007). Bt cotton performance and constraints in central India. *Outlook on Agriculture, 36*(3), 175–180.

Raynolds, L. T. (2000). Re-embedding global agriculture: The international organic and fair trade movements. *Agriculture and Human Values, 17*(3), 297–309.

Raynolds, L. T. (2004). The globalization of organic agro-food networks. *World Development, 32*(5), 725–743.

Raynolds, L. T. (2008). The organic agro-export boom in the Dominican Republic: Maintaining tradition or fostering transformation? *Latin American Research Review, 43*(1), 162–184.

Raynolds, L. T., and Bennett, E. A. (2015). Introduction to research on fair trade. In L. T. Raynolds and E. A. Bennett (Eds.), *Handbook of research on fair trade* (pp. 3–23). Cheltenham, UK: Edward Elgar.

Raynolds, L. T., Murray, D., and Heller, A. (2007). Regulating sustainability in the coffee sector. *Agriculture and Human Values, 24*(2), 147–163.

Raynolds, L. T., Murray, D., and Taylor, P. L. (2004). Fair trade coffee: Building producer capacity via global networks. *Journal of International Development, 16*(8), 1109–1121.

Raynolds, L. T., Murray, D., and Wilkinson, J. (Eds.). (2007). *Fair trade: The challenges of transforming globalization*. Oxon, UK: Routledge.

Reece, W. (2006). Crime, bio-agriculture and the exploitation of hunger. *The British Journal of Criminology, 46*(1), 26–45.

Renard, M-C., and Perez-Grovas, V. (2007). Fair trade coffee in Mexico: At the center of debates. In L. T. Reynolds, D. L. Murray, and J. Wilkinson (Eds.), *Fair trade: The challenges of transforming globalization* (pp. 138–156). London: Routledge.

Reyes, C. M. (2002). *Impact of agrarian reform on poverty*. Manila: Philippine Institute for Development Studies.

Rice Fund Surin (Rice Fund Surin Organic Agriculture Cooperative). (2012). *Raai ngaan kaan damnoen ngaan pii 2554* [Annual report, 2011]. Surin, Thailand: Rice Fund Surin Organic Agriculture Cooperative.

Rigg, J. (2005). Poverty and livelihoods after full-time farming: A South-East Asian view. *Asia Pacific Viewpoint, 46*(2), 173–184.

Rigg, J. (2006). Land, farming, livelihoods, and poverty: Rethinking the links in the rural South. *World Development, 34*(1), 180–202.

Roitner-Schobesberger, B., Darnhofer, I., Somsook, S., and Vogl, C. R. (2008). Consumer perceptions of organic foods in Bangkok, Thailand. *Food Policy, 33*(2), 112–121.

Rola, A., and Coxhead, I. (2002). Does nonfarm job growth encourage or retard soil conservation in Philippine Uplands? *Philippine Journal of Development, 29*(1), 55–83.

Ronchi, L. (2002). *The impact of fair trade on producers and their organizations: A case study with Coocafe in Costa Rica* (PRUS Working Paper No. 11). Brighton, UK: Poverty Research Unit, University of Sussex.

Rowlands, J. (1995). Empowerment examined. *Development in Practice, 5*(2), 101–107.

Roy, D. (2010). Of choices and dilemmas: Bt cotton and self-identified organic cotton farmers in Gujarat. *Asian Biotechnology and Development Review, 12*(1), 51–79.

Roy, D., Herring, R. J., and Geisler, C. C. (2007). Naturalising transgenics: Official seeds, loose seeds and risk in the decision matrix of Gujarati cotton farmers. *Journal of Development Studies, 43*(1), 158–176.

Ruben, R. (Ed.). (2008). *The impact of fair trade*. Wageningen, The Netherlands: Wageningen Academic Publishers.

Ruben, R., and van Schendel, L. (2008). The impact of fair trade in banana plantations in Ghana: Income, ownership and livelihoods of banana workers.

In R. Ruben (Ed.), *The impact of fair trade* (pp. 137–153). Wageningen, The Netherlands: Wageningen Academic Publishers.

Rutten, R. (2010). Who shall benefit? Conflicts among the landless poor in a Philippine agrarian reform programme. *Asian Journal of Social Science, 38,* 204–219.

Sahota, A. (2007). The international market for organic and fair trade food and drink. In S. Wright and D. McCrea (Eds.), *The handbook of organic and fair trade food marketing* (pp. 1–28). Oxford, UK: Blackwell Publishing.

Sanderson, S. (2005). Poverty and conservation: The new century's 'Peasant Question'? *World Development, 33*(2), 323–332.

Sano, D., and Prabhakar, S. V. R. K. (2010). Some policy suggestions for promoting organic agriculture in Asia. *Journal of Sustainable Agriculture, 34*(1), 80–98.

Schmitt, V. G. H. (2012). The partnership between an entrepreneurship support agency and an association: From conception to integration in fair trade. *Journal of International Development, 24*(3), 387–395.

Scialabba, N. E., and Hattam, C. (2002). *Organic agriculture, environment and food security* (Environment and Natural Resources Series). Rome: FAO.

Scoones, I. (1998). *Sustainable rural livelihoods: A framework for analysis* (IDS Working Paper 72). Brighton, UK: Institute of Development Studies.

Scoones, I. (2002). Can agricultural biotechnology be pro-poor? A sceptical look at the emerging consensus. *IDS Bulletin, 33*(4), 114–119.

Scoones, I. (2008). Mobilizing against GM crops in India, South Africa and Brazil. In S. M. Borras Jr., M. Edelman, and C. Kay (Eds.), *Transnational agrarian movements confronting globalization* (pp. 147–176). West Sussex, UK: Wiley-Blackwell.

Scott, J. C. (1972a). Patron–client politics and political change in Southeast Asia. *American Political Science Review, 66*(1), 91–113.

Scott, J. C. (1972b). The erosion of patron–client bonds and social change in rural Southeast Asia. *Journal of Asian Studies, 32*(1), 5–37.

Seri, P. (1988). *Religion in a changing society.* Hong Kong: Arena Press.

Shah, E. (2005). Local and global elites join hands: Development and diffusion of Bt cotton technology in Gujarat. *Economic and Political Weekly, 40*(43), 4629–4639.

Sharma, K. R., and Das, T. C. (2009). *Globalization and plantation workers in North-East India.* Delhi, India: Kalpaz Publications.

Shigetomi, S. (2010). The social investment fund of Thailand: New intermediaries for local development. In J. Midgley and K. Tang (Eds.), *Social policy and poverty in East Asia: The role of social security* (pp. 155–166). New York: Routledge.

Shiva, V. (1991). *The violence of the green revolution.* London: Zed Books.

Siddique, M. A. B. (1995). The labour market and the growth of the tea industry in India: 1840–1900. *South Asia, 18*(1), 83–113.

Sidwell, M. (2008). *Unfair trade.* London: Adam Smith Institute.

Singh, S. N., Narain, A., and Kumar, P. (2006). *Socio-economic and political problems of tea garden workers.* New Delhi, India: Mittal Publications.

Smith, A. M. (2009). Fairtrade, diversification and structural change: Towards a broader theoretical framework of analysis. *Oxford Development Studies, 37*(4), 457–478.

Smith, E., and Marsden, T. (2004). Exploring the "limits to growth" in UK organics: Beyond the statistical image. *Journal of Rural Studies, 20,* 345–357.

Somchai, P. (2006). *Civil society and democratization: Social movements in Northeast Thailand.* Copenhagen, Denmark: NIAS Press.

Somporn, I., and Fukui, S. (2005). Export potentials and constraints for development of jasmine rice production in Thailand. *Journal of International Cooperation Studies, 13*(2), 29–48.

Srinivas, K. R. (2002). Bt cotton in India: Economic factors versus environmental concerns. *Environmental Politics, 11*(2), 159–164.

Srinivasan, C. S. (2003). Concentration in ownership of plant variety rights: Some implications for developing countries. *Food Policy, 28*(5–6), 519–546.

Srivastava, V. K. (2008). Concept of "tribe" in the draft national tribal policy. *Economic and Political Weekly, XLIII*(50), 29–35.

Stevis, D. (2015). Global labor politics and fair trade. In L. T. Raynolds and E. A. Bennett (Eds.), *Handbook of research on fair trade* (pp. 102–119). Cheltenham, UK: Edward Elgar.

Stone, G. D. (2007). Agricultural deskilling and the spread of genetically modified cotton in Wagangal. *Current Anthropology, 48*(1), 67–103.

Subramanian, A., and Qaim, M. (2009). Village-wide effects of agricultural biotechnology: The case of Bt cotton in India. *World Development, 37*(1), 256–267.

Subramanian, A., and Qaim, M. (2010). The impact of Bt cotton on poor households in rural India. *Journal of Development Studies, 46*(2), 295–311.

Sundaram, K., and Tendulkar, S. D. (2003). Poverty among social and economic groups in India in 1990s. *Economic and Political Weekly, 38*(50), 5263–5276.

Sununtar, S., Leung, P., and Cai, J. (2006). *Contract farming and poverty reduction: The case of organic rice contract farming in Thailand* (ADB Institute Discussion Paper No. 49). Manila: Asian Development Bank.

Tallontire, A. (2000). Partnership in fair trade: Reflections from a case study of cafedirect. *Development in Practice, 10*(2), 166–177.

Taylor, J. (1997). "Thamma-chaat": Activist monks and competing discourses of nature and nation in northeastern Thailand. In P. Hirsch (Ed.), *Seeing forests for trees: Environment and environmentalism in Thailand* (pp. 37–52). Chang Mai, Thailand: Silkworm Books.

Thapa, U., and Tripathy, P. (2006). *Organic farming in India: Problems and prospects.* Udaipur, India: Agrotech Publishing Academy.

Thavat, M. (2011). The tyranny of taste: The case of organic rice in Cambodia. *Asia Pacific Viewpoint, 52*(3), 285–298.

Thiers, P. (2005). Using global organic markets to pay for ecologically based agricultural development in China. *Agriculture and Human Values, 22*(1), 3–15.

Thunya, S. (2010). *Khaya im: Thaang lueak khoong kaan catkaan singwaetloom phaak prachaachon* [Smiley garbage project: An alternative way of environmental management]. Surin, Thailand: Community of Agro-ecology Foundation.

Tovey, H. (1997). Food, environmentalism and rural sociology: On the organic farming movement in Ireland. *Sociologia Ruralis, 37*(1), 21–37.

Triiyadaa, T. (2012). *Talaat nat sii khiaw mueang Surin* [Regular green market in Surin City]. Surin, Thailand: Community of Agro-ecology Foundation.

Tripp, R. (2001). Can biotechnology reach the poor? The adequacy of information and seed delivery. *Food Policy, 26*(3), 249–264.

Tripp, R. (Ed.). (2009). *Biotechnology and agricultural development: Transgenic cotton, rural institutions and resource-poor farmers.* Oxon, UK: Routledge.

Truscott, L., Lizarraga, A., Nagarajan, P., Tovignan, S., and Currin, A. (2010). *2010 farm and fiber report: Organic by choice.* O'Donnell, TX: Textile Exchange.

Truscott, L., Lizarraga, A., Nagarajan, P., Tovignan, S., and Denes, H. (2011). *2011 midyear farm and fiber predictions report: Organic by choice*. O'Donnell, TX: Textile Exchange.

Tsuruta, T., and Suriya, C. (2016). A preliminary report on diversity of products in organic farmers' markets in Surin, Northeast Thailand. *Memoirs of Faculty of Agriculture, Kinki University, 49*, 67–80.

TV Burabha. (2010). *Khrueakhaai khon kin khaaw kueakuun chaawnaa* [Network of rice-eaters to support paddy growers]. Retrieved November 30, 2014, from http://www.tvburabha.com/tvb/rice/ta2.html

Tyner, J. A., and Donaldson, D. (1999). The geography of Philippine labour international migration fields. *Asia Pacific Viewpoint, 40*(3), 217–234.

Utting, K. (2009). Assessing the impact of fair trade coffee: Towards an integrative framework. *Journal of Business Ethics, 86*, 127–149.

Vacher, C., Bourguet, D., Rousset, F., Chevillon, C., and Hochberg, M. E. (2004). High dose refuge strategies and genetically modified crops – reply to Tabashnik et al. *Journal of Evolutionary Biology, 17*(4), 913–918.

Valkila, J., and Nygren, A. (2010). Impacts of fair trade certification on coffee farmers, co-operatives, and Labourers in Nicaragua. *Agriculture and Human Values, 27*(3), 321–333.

Vandergeest, P. (2009). *Opening the green box: How organic became the standard for alternative agriculture in Thailand*. Paper presented at the Berkeley Workshop on Environmental Politics, University of California, Berkeley, 17 April 2009.

Vitoon, P. (Ed.). (2001). *Maatathaan kaseet insii chabap kaatuun* [Standard of organic agriculture: Cartoon version]. Bangkok: Green Net.

Vitoon, P. (2012). Overview of Thai organic agriculture. In H. Willer and L. Kilcher (Eds.), *The world of organic agriculture: Statistics and emerging trends 2012* [FiBL-IFOAM report] (pp. 190–192). Bonn, Germany: FiBL and IFOAM.

Vitoon, P. (2015). *Phaap ruam kaseet insii thai 2556–57* [A general outline of Thai organic agriculture, 2013–14]. Bangkok: Green Net and Earth Net Foundation.

Wai, O. K. (2012). Organic Asia 2012. In H. Willer and L. Kilcher (Eds.), *The world of organic agriculture: Statistics and emerging trends 2012* (pp. 170–177). Bonn, Germany: Research Institute of Organic Agriculture (FiBL), Frick, and International Federation of Organic Agriculture Movements (IFOAM).

Walaiporn, O., Areerat, K., and Manitchara, T. (2007). *Organic and inorganic rice production: A case study in Yasothon Province, Northeast Thailand*. Penang, Malaysia: Pesticide Action Network Asia and the Pacific.

Weale, A. (2010). Ethical arguments relevant to the use of GM crops. *New Biotechnology, 27*(5), 582–587.

Weber, J. (2007). Fair trade coffee enthusiasts should confront reality. *Cato Journal, 27*(1), 109–117.

Wilkinson, J., and Mascarenhas, G. (2007). Southern social movements and fair trade. In L. Raynolds, D. Murray, and J. Wilkinson (Eds.), *Fair trade: The challenges of transforming globalization* (pp. 125–137). London: Routledge.

Willer, H., and Kilcher, L. (Eds.). (2012). *The world of organic agriculture: Statistics and emerging trends 2012*. Bonn, Germany: FiBL and IFOAM.

Wilson, G. A., and Rigg, J. (2003). "Post-productivist" agricultural regimes and the South: Discordant concepts? *Progress in Human Geography, 27*(6), 681–707.

Withuun, L. (Ed.). (1996). *Kaseettrakam thaang lueak: Khwaam maai khwaam pen-maa lae teknik wiithii* [Alternative agriculture: Its meaning, history, and technology]. Bangkok: Alternative Agriculture Network.

Woods, M. (2011). *Rural*. London: Routledge.

World Bank. (2009). *Land reform, rural development and poverty in the Philippines: Revising the agenda*. Pasig City, Philippines: World Bank.

World Fair Trade Organization (WFTO) and Fairtrade International (FLO). (2009). *A charter of fair trade principles*. Retrieved December 30, 2015, from http://wfto.com/fair-trade/charter-fair-trade-principles

World Intellectual Property Organization. (2016). *Geographical indications*. Retrieved May 21, 2016, from http://www.wipo.int/geo_indications/en/

Wyatt, B. (2010). Local organic certification in northern Thailand: The role of discourse coalitions in actor-networks. *International Journal of Sociology of Agriculture and Food, 17*(2), 108–121.

Wynen, E. (2003). What are the key issues faced by organic producers? In Organisation for Economic Co-operation and Development (Ed.), *Organic agriculture: Sustainability, markets and policies* (pp. 207–220). Oxfordshire, UK: Cabi Publishing.

Zerbe, N. (2004). Feeding the famine? American food aid and the GMO debate in Southern Africa. *Food Policy, 29*(6), 593–608.

Index

Note: figures and tables are denoted with italicized page numbers. End note information is denoted by an n and note number following the page number.